PENGUIN BOOKS

THE NUMBERS GAME

'A fascinating and stylish investigation into a rapidly developing way of understanding football'
Jonathan Wilson, author of *Inverting the Pyramid: The History of Football Tactics*

'Whether you are a traditionalist or a numbers nut you can enjoy this book. It's thorough, accessible and devoid of the absolute truths so many on both sides of the debate peddle'
Gabriele Marcotti, football broadcaster and author

'Pleasingly counter-intuitive . . . I learnt a lot. It's hard not to applaud a book that is bent on the disenchantment of football's internal conversations and archaic practices, while simultaneously acknowledging an ineradicable core of the unpredictable and random at its heart'
David Goldblatt, author of *The Ball is Round: A Global History of Football*

'Be warned: *The Numbers Game* will change the way you think about your favourite team or player, and change the way you watch the beautiful game'
Billy Beane, General Manager of the Oakland A's, the subject of *Moneyball*

'I couldn't wait to get hold of a copy. An important step in the development of the popular football literature'
Stefan Szymanski, co-author of *Soccernomics*

'Engaging and stimulating. *The Numbers Game* overturns several tenets of football thinking [and is] a valuable addition to the scarce literature at a time when pioneers inside football are only just starting to work out which stats matter, while people outside the game still scarcely know that anything is changing. It's also energetically and cleanly written, and is free of academic jargon' Simon Kuper, *New Statesman*

'A must read' *Tompkins Times*

'A fascinating book. Some of the facts are astounding'
Alison Kervin, sports editor, *Mail on Sunday*

'A book that makes you question what you thought you knew' *Sky Sports*

'Debunk(s) the myths and conventional wisdom that have defined soccer for the last century' Jack Bell, *The New York Times*

'It will not only help [fans] understand the game better, but it will also stimulate new ways to analyze and think about the game'
Zach Slaton, *Forbes*

'I may return to this book (likelihood high, at least 90 per cent)' *Irish News*

'Do yourself a favour and buy this book' *Irish Examiner*

'A great piece of work' Damien O'Meara, *Game On* on RTÉ R2

'Mind-engaging, mind-stimulating, and even a bit mind-expanding'
SB Nation

'The best sports book of the summer' Simon Mayo, BBC Radio 2

ABOUT THE AUTHORS

At seventeen, Chris Anderson found himself playing in goal for a fourth division club in West Germany; today, he's a professor in the Ivy League at Cornell University in Ithaca, New York. An award-winning social scientist and football analytics pioneer, Anderson consults with leading clubs about how best to play the numbers game.

David Sally is a former baseball pitcher and a professor at the Tuck School of Business at Dartmouth College in the US, where he analyses the strategies and tactics people use when they play, compete, negotiate and make decisions. He is an advisor to clubs and other organizations in the global football industry.

The Numbers Game

Why Everything You Know About Football is Wrong

CHRIS ANDERSON AND
DAVID SALLY

PENGUIN BOOKS

PENGUIN BOOKS

Published by the Penguin Group
Penguin Books Ltd, 80 Strand, London WC2R ORL, England
Penguin Group (USA) Inc., 375 Hudson Street, New York, New York 10014, USA
Penguin Group (Canada), 90 Eglinton Avenue East, Suite 700, Toronto, Ontario, Canada M4P 2Y3
(a division of Pearson Penguin Canada Inc.)
Penguin Ireland, 25 St Stephen's Green, Dublin 2, Ireland (a division of Penguin Books Ltd)
Penguin Group (Australia), 707 Collins Street, Melbourne, Victoria 3008, Australia
(a division of Pearson Australia Group Pty Ltd)
Penguin Books India Pvt Ltd, 11 Community Centre, Panchsheel Park, New Delhi – 110 017, India
Penguin Group (NZ), 67 Apollo Drive, Rosedale, Auckland 0632, New Zealand
(a division of Pearson New Zealand Ltd)
Penguin Books (South Africa) (Pty) Ltd, Block D, Rosebank Office Park,
181 Jan Smuts Avenue, Parktown North, Gauteng 2193, South Africa

Penguin Books Ltd, Registered Offices: 80 Strand, London WC2R ORL, England

www.penguin.com

First published by Viking 2013
Published with a new chapter, 'Extra Time – The Numbers Game
at the World Cup', in Penguin Books 2014
002

Copyright © Chris Anderson and David Sally, 2013, 2014
All rights reserved

The moral right of the authors has been asserted

Typeset by Jouve (UK), Milton Keynes
Printed in Great Britain by Clays Ltd, St Ives plc

Except in the United States of America, this book is sold subject
to the condition that it shall not, by way of trade or otherwise, be lent,
re-sold, hired out, or otherwise circulated without the publisher's
prior consent in any form of binding or cover other than that in
which it is published and without a similar condition including this
condition being imposed on the subsequent purchaser

ISBN: 978–0–241–96362–3

www.greenpenguin.co.uk

Penguin Books is committed to a sustainable
future for our business, our readers and our planet.
This book is made from Forest Stewardship
Council™ certified paper.

To our home teams
Kathleen, Nick and Eli
Serena, Ben, Mike, Tom and Rachel

Contents

Football for Sceptics – The Counter(s) Reformation

In sports, what is true is more powerful than what you believe, because what is true will give you an edge.

Bill James

Seven words have long dominated football:

That's the way it's always been done.

The beautiful game is steeped in tradition. The beautiful game clings to its dogmas and its truisms, its beliefs and its credos. The beautiful game is run by men who do not wish to see their power challenged by outsiders, who know that their way of seeing the game is the true way of seeing the game. They do not want to be told that, for more than a century, they have been missing something. That there is knowledge that they do not possess. That how things have always been done is not how things *should* always be done.

The beautiful game is wilful in its ignorance. The beautiful game is a game ripe for change.

And at the centre of that change are numbers. It is numbers that will challenge convention and invert norms, overhaul practices and shatter beliefs. It is numbers that let us glimpse the game as we have never seen it before.

Every world-class club knows this. All of them employ analysis staff, specialists in data collection and interpretation who use all the information they can glean to plan training sessions, design playing systems, plot transfers. There are millions of pounds and hundreds of trophies at stake. Every club is prepared to do anything it takes to gain the slightest edge.

But what none of those clubs has yet managed to do is take those numbers and see their inner truth. It is not just a matter of collecting data. You have to know what to do with them.

This is football's newest frontier. It is often said that football cannot, or should not, be broken down into mere statistics. That, critics say, removes the beauty from the beautiful game. But that is not how the clubs who fight to win the Champions League or the Premier League or the nations battling to lift the World Cup see it, and neither do we. We believe that every shred of knowledge we can gather helps us love football, in all of its complex glory, all the more. This is the future. There is no stopping it.

That is not to say all of football's traditions are wrong. The data we are now able to gather and analyse confirm that some of what we've always thought was true really is true. Beyond this, however, the numbers offer us further truths, make clear things we could not have known intuitively and expose the falsehoods of 'the way it's always been done'. The biggest problem resulting from following a venerated tradition and hardened dogma is that they are rarely questioned. Knowledge remains static while the game itself and the world around it change.

Asking Questions

It was a simple question, asked in that bewildered tone Americans often use when discussing football.

'Why do they do that?'

Dave and I were watching Premier League highlights, and something had caught his eye. Not a moment of dazzling skill, or bewitching beauty, or even inept refereeing, but something altogether more mundane. Dave was baffled, like countless central defenders before him, by Rory Delap's long throws.

Every single time Stoke City won a throw-in within hurling distance of the opposition box, Delap would trot across to the touchline, dry the ball with his shirt – or, when at home, with a towel handily placed for that very purpose – and proceed to catapult it into the box, over and over and over again.

To me, as a former goalkeeper, the benefits of Delap's throws were obvious. I explained it to Dave: Stoke had a decent team, but one lacking a little in pace and even more in finesse. What they did have, though, was height. So why not, when the ball goes out of play, take the opportunity to create a chance out of nothing? Why not cause a little havoc in your opponents' ranks? It seemed to work.

That did not sate Dave's curiosity, though. It simply served to make him ask the next logical question.

'So why doesn't everyone do it?'

The answer to that was equally obvious: not everyone has a Rory Delap, someone capable of hurling the ball great distances with that flat trajectory, like a skimmed stone, that panics defenders and confuses goalkeepers.

Dave, himself a former baseball pitcher, tried another tack: 'But can't you try and find one? Or make one of your players lift weights and practise the javelin and the hammer?'

There was a problem with this. Yes, Dave's questions, like those of a persistently inquisitive child, were getting annoying; more irritating still, I did not have a good answer.

'You could play the game the way Stoke do,' I countered, 'if you have a Delap and loads of tall central defenders. But it's just not very attractive. It's not what you do unless you have to.'

'Why?' Dave responded, with crushing logic. 'It seems to work for them.'

And that was it. All I had left, like a frustrated parent, was one word. 'Because.'

Because there are some things you don't want to do when playing football. Because, even though a goal created by a long throw is worth just as much as one from a flowing passing move, it's almost like it doesn't count as much. Because, to a purist, they're somehow not quite as deserved.

But Dave's endless questions – Why? Why? Why? – nagged at me. If it works for Stoke, why don't more teams do it? Who was right? Stoke, who were responsible for almost a third of all the goal-scoring chances from throw-ins created in the Premier League that year – or everyone else, who clearly felt they did not need, or did not want, the long throw in their arsenal?

Why are there some things that are just 'not done'?

Why is football played the way it is?

We attempted to answer these two very big questions by applying our knowledge and skills – as a political economist in my case and a behavioural economist in Dave's – our discipline as social scientists, our experiences as a goalkeeper and a baseball pitcher, and our love for sports and for solving hard problems. The result rests in your hands – a book about football and numbers.

Football has always been a numbers game: 1–1, 4–4–2, the big number 9, the sacred number 10. That will not change and we don't ever want it to. But there is a 'counters-reformation' gathering pace that may make another set of figures seem just as important: 2.66, 50/50, 53.4, <58<73<79, and 0 > 1 will all prove to be essential for the future of football.

This is a book about football's essences – goals, randomness, tactics, attack and defence, possession, superstars and weak links, development and training, red cards and substitutions, effective leadership, and firing and hiring the manager – and the way these relate to numbers.

The Analytics Hub

The neat, unassuming, thoughtful types who make their way to Boston every March for the Sports Analytics Conference hosted by the prestigious MIT Sloan School of Management make unlikely gurus for anyone seeking a glimpse into football's future or its essence. But these are the coaches, staff and executives of the world's major sports teams who gather every year to develop, learn about and map out the numbers game.

Association football is a sport that has long been determined by finely tuned athletes and stony-faced managers. The sort of men and women who would sit happily through such presentations as 'Deconstructing the Rebound with Optical Tracking Data', or the irresistible 'Next Generation Sports Coaching Using Mobile Devices', have not found the game a welcoming environment. That, though, is starting to change. Analytics – 'the discovery and communication of meaningful patterns in data'[1] – is booming in dozens of industries, and sport is beginning to awaken to its potential. Analytics is much more than just spreadsheets and statistics: it is an openness to data and information of all kinds – formal, informal, categorized, disorganized, observed, recorded, remembered, etc. – and it is a determination to find whatever truth, patterns and correspondences they may contain. Baseball, basketball and American football have embraced analytics. Football is some way behind, more reluctant to embrace the future.

Among the 2,000 or so delegates – up from 200 in 2007 – are representatives of some of Europe's leading football clubs, as well as the data-producing companies who try to quench the game's seemingly insatiable thirst for information.

They are, for now, just a handful – delegates from the US sports still form the core audience; David Gill, Chief Executive of Manchester United, can wander these halls unmolested, as he did at the 2012 conference, while Bill James, baseball's analytics pioneer, is treated as a celebrity – but they are increasing in number every year.

Analytics is sport's cutting edge, and in football it is growing exponentially. Managers, scouts, players and owners all want an advantage, and knowledge is power. These are the men and women who supply it. Every year, in that Boston convention centre, gather the game's new pioneers.

They are not there simply to discuss how to collect as much data as possible. As Albert Einstein said: 'Not everything that counts can be counted, and not everything that can be counted counts.' Instead, they want to know how they can use that data to win this week, this season. That's not an easy task. Clubs are being inundated by a torrent of information as the nascent science of analysis explores its possibilities. Mike Forde, Chelsea's forward-thinking Director of Football Operations, claims that his team has gathered around '32 million data points from something like 12,000 or 13,000 games'.[2]

Some of these will have been gathered by the club itself, from scouting and match reports, recorded on the state-of-the-art video and computer equipment no self-respecting football club would be without. The rest of it will have been provided by the likes of Opta, Amisco, Prozone, Match Analysis, or StatDNA, companies that provide clubs with ever more elaborate data sets to pore over in search of the slightest gain. Aside from match data, clubs also keep detailed medical records and training logs – injury prevention and rehabilitation are among the frontiers in football analytics – along with data on which players sell more shirts and which put more bums on more seats and which games sell more pies and pints. There is an arms race here: clubs and companies desperate to outdo each other to prove how comprehensive they can be, how many things they can count.

Gathering the information is just the first step. The clue to analytics is in the name. To make those numbers mean something, to learn something from them, they must be analysed. The key, for those at the vanguard of what some have called a data 'revolution' and what we think of as football's reformation, is to work out what they need to be counting, and to discover why, exactly, what they are counting counts.

Football Analytics Today

Deep inside Roberto Martínez's home stands a 60-inch, pen-touch television screen; it is linked to his personal computer, which is loaded with Prozone's most advanced software. After returning from a match, Everton's Spanish manager – who will emerge as one of the heroes of this book – will spend hours locked away watching his side's latest game again and again; often, he will need to see the fixture ten times before he is satisfied. 'My wife was delighted when I had it installed,' Martínez told the *Daily Mail*. 'But she understands I need that space and time to be able to come back to being myself. Once I find a solution, I'm fine.'[3]

Martínez is far from atypical. Football can still be an old-fashioned business, where managers follow the time-honoured tradition of collecting intelligence and information by themselves, by watching players in training and in matches, reading the news, consulting their staff, listening to their scouts. But clubs at the elite level complement that with an analysis department, staffed by trusted adjutants who can help their manager see what is and isn't there.

That is what Steve Brown and Paul Graley do for Martínez at Everton. The manager's match analysts spend hours preparing and auditing Premier League games in meticulous detail, examining the attack and defence of their own players and the opposition, preparing background materials on each player's immediate opponent. Before a match, they will examine at least five of the opposition's previous games, compiling scouting reports and combining them with Prozone's data. Using these data and video, they look at style, approach, strengths, weaknesses, positional organization and the follies and foibles of their players. All of that is boiled down and presented to Martínez, who summarizes further and delivers the assessment to his squad.

Brown and Graley also work one-on-one with individual players. Some will sit with them before the match to do some homework and go through their direct opponent's patterns of play. Sometimes they cram collectively, talking things through as late as game day – especially when opposition players are either playing in unusual positions or new players come into the opposition team. As soon as the match is over, the

Everton staff start their post-mortem. Graley will go through the game a number of times, along with the coaches, and summarize and review what worked and what didn't. Again, the manager is part of the process, and individual players regularly learn what they did well and badly so they can adjust for next time.

You might think that the men whose job it is to break down your team's and your opponent's strengths and weaknesses – the men who hold the key to next Saturday's victory – sit near the centre of Everton's universe, right next door to the manager.

And yet, when we visited them at the club's Finch Farm base on the outskirts of Liverpool, we found that their office is just one of many along a corridor leading to the canteen. It is a functional, unspectacular space. There are few clues as to the nature of the work that goes on here: file folders sit on top of standard-issue desks next to desktop computers; Steve and Paul sit on ordinary swivel chairs. It could be any office, anywhere, in any industry.

Only the tactics whiteboard in the corner, and the software on screen, hint that this is a room dedicated to analysing the best way to maximize performance in one of the world's most glamorous, rich and exciting leagues.

It is somehow fitting that the analysts at Everton – and those we've seen elsewhere – are but one spoke in the wheel of a club's football operation. Brown and Graley and their ilk are relatively novel creatures. Generally, in football, nobody is quite sure what to do with them. They are the latest addition to the manager's back-room staff; not as established as coaches, scouts, physiotherapists or even psychologists, their place in the pecking order is uncertain.

Their arrival, though, has not gone unnoticed by the market.

In the decade or two since the first football analysts were appointed a whole industry of data providers has emerged to satisfy their appetite, their endless desire, for more – and better – information to pass on to their managers.

The first of these companies to emerge was Opta Sports, started by a group of management consultants who, in the 1990s, decided to create an index of player performance in football. As Content Director Rob Bateman told us, the aim was simply 'to get the brand into the public eye'. Opta contacted the Premiership (as the top tier of English football was known between 1993 and 2007); they were given funding by Carling, who sponsored the league at the time, and former Arsenal and England coach Don Howe came on board to provide football expertise. They launched the index in 1996 on Sky Sports and in the *Observer* newspaper, but soon discovered that the information they were collecting was far more valuable than the publicity the index brought the company. They could sell it to media outlets, near and far; later, they would discover that clubs were just as desperate for it.

When Opta started, each game's events took about four hours to code, using a pen and paper and pressing stop/start on a video recorder. The actions they noted were basic: passes, shots, saves. The level of detail their analysts record now is a world away from those unassuming beginnings. Take the 2010 Champions League final between Bayern Munich and Inter Milan. That night, Opta's team of three analysts logged a total of 2,842 events, around one every two seconds of the game. One was designated to monitor Inter, one Bayern, each one an expert in their subjects – they had been following their games, tracking all their actions and movements, all season. They were joined by a teammate in the role of overseer, pointing out mistakes and omissions.

More than a decade on from their birth, though, Opta are just one of a number of path-breaking companies formed to satisfy football's increasing addiction to data. Everton, as we saw when we were welcomed into Steve Brown's inner sanctum, subscribe to Prozone, a Leeds-based company set up to deliver data specifically designed to help with the coaching and scouting of players. In summer 2011 it merged with a French rival, Amisco, and between them the two brands now are among the industry's leaders.

Where clubs had once relied on good relations with their opponents to obtain videos of their most recent games – a system dependent on reciprocal trust which often proved misplaced when match videos were inexplicably lost – Amisco and Prozone developed the technology not only to allow the rapid analysis of a team's matches, but to collect even more data.

They mounted cameras high above the pitch to track individual players, to give coaches, sports scientists and the like the sort of information they craved: how much running a player did and at what speed, how the flow of the game affected events. Later they combined the video with software that allowed players and actions to be tagged: now it is easy to compile footage of an individual's actions, or of all the goals your opponents have conceded. Martínez can watch all his team's corners or all his midfield's misplaced passes from a comfy armchair at home at the click of a button.

Prozone and Opta are not alone. There are many other companies working in the same arena across the world: Impire in Germany, Infostrada in the Netherlands and Match Analysis and StatDNA in the United States . . .

All are benefiting from the boom as the markets to which they sell their data expand seemingly without limit. There are the coaches, players, executives, journalists, fans and even academics who have a growing appetite for football's numbers, and then the video game manufacturers, fantasy football leagues and the betting houses which use them to make money.

Those involved in assessing, managing and exploiting risk, whether it be in financial markets or sports gambling, tend to build elaborate forecasting models. For that, they need data. Bookmakers' odds are not set on a whim; all the data they can access is fed into one of their algorithmic engines, and favourites and outsiders are determined accordingly. Algorithms are equally key when determining prices on the financial markets. Football is right at the intersection of the two areas.

Just as the betting companies are raking in the profits from their analytical, odds-setting engines – and using them to fund expensive sponsorship deals with the biggest names in sport, such as bwin's current arrangement with Real Madrid – those men who made their fortunes playing the markets are buying into the game: Sunderland, Brentford, Brighton, Stoke, Liverpool, Millwall and many others all have owners who do not place a bet or invest a penny without examining the numbers first.

That is the true power of data: to change our relationship with the game. Owners no longer have to rely on their own judgment to discern whether their team is performing well or if their investment is sound – the numbers can be slipped on to their desk every Monday, or even sent Sunday morning to their mobiles or iPads. After every training session managers can post data on the dressing-room door showing how far each player ran.

And some of that information is available to fans, published in newspapers or flashed up on the television screen, available at the push of a button on a smartphone and recorded for ever online. There is no hiding place. The eye in the sky is always watching. No wonder Paul Barber, formerly a director at Tottenham Hotspur and now Chief Executive of Brighton and Hove Albion, refers to the rise in and increased sophistication of video analysis as being 'like an X-Ray'.[4] This is the age of the see-through footballer: it is little surprise that the game's radiographers – men like Steve Brown and Paul Graley – are finding themselves slowly, incrementally, welcomed in from the cold.

The days of relying purely on gut instinct, conjecture and tradition to judge what constitutes good and bad football are over; instead, we can turn to objective proof. The implications are profound. The use of objective information is reshuffling the balance of power in the beautiful game. Instead of being run by a mix of command, habit and guesswork, football is entering a new, more meritocratic phase.

That is threatening to the game's traditional power brokers, because it suggests there may be something they have been missing all these years. In that sense football is a little like a religion: there has long been a perception that, to be an expert, you must have been born in the right place and been steeped in its rituals from a young age. There are creeds, dogmas, communion with your fellow fans, confessions, dress codes, imbibing and chanting and all the rest.

But if the data allow just anyone to become an expert, to have an informed opinion, those immersed in the old ways become less powerful, less special, more open to question. Ultimately they can be proved wrong, and the more they are proved wrong the less power they have. If they are the priests and the papists, our role as authors of *The Numbers Game* is to teach you both to be and to appreciate the iconoclasts and counters of football's reformation.

This, perhaps, explains the degree of resistance football's analytics pioneers have encountered.

We were tasked by one club, before one recent transfer window, with a research project that focused on strengthening their squad in particular areas. We were delighted to hear that our results had been received well by the board. The manager, though, was rather less enthusiastic. 'Stats can't tell me who to sign,' he said. 'They can't measure the size of a player's heart.'

It is the same with using data to adapt your approach to a particular match. 'The manager believes it when he sees it with his own eyes,' one Premier League match analyst told us. 'He likes to watch the video, and he tries to go out and see as many matches as he can for himself.'

This is not only an English problem; reluctance to embrace new technology, new sources of information, spreads far and wide.

Boris Notzon, the Director of 1. FC Köln's SportLab, showed us around one of the most advanced analytics enterprises in professional football. Köln employs three full-time and thirty part-time analysts from fifteen countries to collect and manage everything from opposition scouting reports to physical data from the club's first, reserve and youth teams. Even he, though, admits that Köln are unusual. As part of a joint project, all German first and second division clubs have access to match data provided by Impire, who use technology similar to that of Opta and Prozone/Amisco. Yet, few actually trust or use the data that are accumulating with every match. They don't want football on a spreadsheet; they want to see it with their own eyes.

'In comparison to historical medicine, football analytics is currently in the time of leeches and blood lettings,' says Mark Brunkhart, founder of Match Analysis. 'Not that we should stop progressing and working, but we should realize how little we understand.'

Football Analytics Yesterday

Football analysts may have become a regular feature of clubs' back-rooms only in recent years, and the technology they use may still be blossoming, but that is not to say that the idea of rigorous game analysis is new. In fact, it's been around for decades.

It would be unfair to describe football's modern-day relationship with analytics as a revolution, but it is somehow more than mere evolution. Perhaps the best word is reformation: the game is the same, but the way it is played is changing. And we are in the most exciting stage of that process, when different aspects emerge every day, every week, every year, when progress is made rapidly, every advance taking things further away from the work of the man who could be considered the first football analyst: Wing Commander Charles Reep.

The Englishman is one of the key characters – the true and tragic hero, in a way – in the story of football analytics. His theories may have been confounded and his beliefs rubbished, but in order to appreciate how far we have to go, we must understand where we have come from.

Reep was not a football man. Born in Cornwall in 1904, he trained as an accountant before joining the Royal Air Force after winning first prize in an entrance competition for the RAF's new Accountancy Division. One evening in 1933 Reep's division was graced by an appearance from Charles Jones, captain of Herbert Chapman's all-conquering Arsenal.

Jones came to give a talk about the club's system of play, and went into detail analysing the understanding that had developed between the right and left wingers in Chapman's side. Reep was agog. He was moved to apply what he knew – accountancy – to what fascinated him – football. And so he set about developing a system for annotating every action that took place on the pitch. The Football Accountant was born.

The point, in Reep's words, was to 'provide a counter to reliance upon memory, tradition and personal impressions that led to speculation and soccer ideologies'.[5] He would deal in facts. He would help us see what we could not see.

Unfortunately, his military career and war intervened, and the first game Reep annotated came on 18 March 1950, fully seventeen years after Jones had visited the RAF's Accountancy Division and fired his imagination. As he watched Swindon face Bristol Rovers, Reep took a pencil and notebook from his pocket and a science was born. 'The continuous action of a game is broken down into a series of discrete on-the-ball events, such as a pass, centre or shot,' said Reep of his system. 'A detailed categorization [is] made for each type of event, for which shorthand codes have been developed. For example each pass in a game is classified and recorded by its length, direction, height and outcome, as well as the positions on the field at which the pass originated and ended.'[6]

Reep was dedicated. He continued attending games well into his nineties, his passion for the sport and the numbers undimmed. He annotated more than 2,200 fixtures over the course of his career, spending around eighty hours analysing each game. Allowing for sleep, that adds up to around thirty years of his life. He would often go to night matches wearing a miner's helmet, complete with headlamp, so he could see his notations. His most magnificent possession was the complete set of notes he made during the 1958 World Cup final, fifty pages of drawings accounting for the movement of the ball throughout the game, all written on a roll of wallpaper.

The data he collected eventually became the basis for a scientific paper – 'Skill and Chance in Association Football' – written with Bernard Benjamin, Chief Statistician at the General Register Office and published in the *Journal of the Royal Statistical Society* in 1968. The aim was to see if the information Reep had painstakingly collected over fifteen years, between 1953 and 1967, revealed predictable patterns in the events of a match.[7]

It was only a short academic paper, but it was powerful. It proved that Reep's coding system did lend itself to scientific analysis. And it showed, for the first time, that several aspects of the game did follow strong and stable numerical patterns. Reep and Benjamin found that, on average, teams scored with roughly one of every nine shots they took. They discovered that a team's odds of completing a pass were generally only as good as a coin toss – around 50 per cent – but that they diminished with each additional pass completed. Football, they determined, was a stochastic (i.e. random) process: one in nine shots yielded a goal, but which one of them would go in was hard to say.

It was also, they discovered, a game of turnovers: the vast majority of movements ended after zero or one completed pass, while 91.5 per cent never reached a fourth successful pass. This distribution of passes was present in most matches that Reep watched, and even today's matches are replete with turnovers. '[In] the average game, the ball changes hands 400 times,' says Chelsea's Mike Forde.[8]

Reep also unearthed another cornerstone of modern football thinking: that 30 per cent of all regained possessions in the opponent's penalty area led to shots on goal; and about half of all goals came from those same regained possessions.

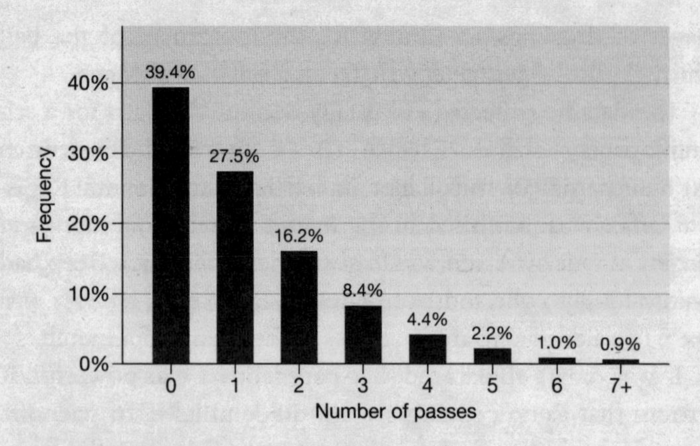

Figure 1 Passing move distributions, 1953–67
Source: *Reep and Benjamin (1968).*

Note: Horizontal axis shows the number of successful passing moves, where 0 means that a pass attempt was immediately intercepted; 1 means one successful pass before possession lost and so on. Numbers atop bars indicate the percentage of moves in a match. Reep and Benjamin found only 8.5 per cent of passing movements contained more than three passes.

When Liverpool signed Stewart Downing and Jordan Henderson in the summer of 2011, more than sixty years after Reep first took pencil from pocket and set his system to work, the pair's 'final third regain' percentage was one of the key statistics used to assess their worth; Barcelona and Spain have based much of their recent success on the pressing game.

Reep did not invent the pressing game, but he was the first to label it; his study offered insights – ways of thinking and talking about the game – not previously imagined.[9] The sport should consider him a pioneer. Instead, he has been condemned as a pariah. Not because he looked at the beautiful game through numbers, but because of what he thought the numbers said.

Confirming Beliefs with Data

Reep was a product of his time. The Football Accountant was not content simply to collect data for his own enjoyment. He saw another use for his findings. Reep had been focused, ever since that visit by Arsenal's Charles Jones to his RAF base, on what it takes to win football matches. To do that, he thought that a team needed to maximize its goal-scoring opportunities. And to do that, he decided, they needed to be as efficient as possible. It was no coincidence that Reep titled his paper, the culmination of his life's work, 'Skill and Chance'. He recognized that football is a game as much reliant on fortune as it is on ability: his discovery that the odds of a pass being completed at any one time are no more than 50/50 is enough to prove that. His aim was to find a way to alter that balance: to use skill to overcome chance.

The solution he hit upon was efficiency. He wanted maximum productivity with minimum wasted effort. This sort of thinking was prevalent when Reep was at his peak. In Britain of the 1940s and 1950s, the stock of accountancy and the faith in data were on the rise, thanks to Keynesian economics, which promised to steer the country's economy by using government spending to manipulate investment and consumption. It was the philosophy introduced to overcome the Great Depression and to survive the rigours of World War II: to do more, with less.

For the principle to work, the government needed data. Good data. And so the Treasury set about collecting statistics on all manner of economic activity: a drive to improve efficiency by compiling data. That, to Reep, was the aim of his football accounting: to beat chance, a team had to be maximally efficient. Teams were more efficient if they scored more goals with less possession, and fewer passes, shots and touches.

Reep had the data to back up his vision, or at least he thought he did. He had proved that only two of every nine goals came from moves in which there were more than three passes. He knew that teams scored once every nine shots; and he knew that a vast portion of goals scored came from regains in and around the opposition's penalty area. And so he concluded, sweepingly and without doubt, that teams were – statistically speaking – better off if they spent less time trying to string together passes and more time moving the ball quickly and efficiently into their opponents' box. And so the efficiency of the long-ball game – minimum input and maximum output – was confirmed.

Being comfortable with numbers isn't the same as producing insights, though. Reep was a superb accountant of the game, but he was no analyst. He failed to ask the analyst's most important question – how might I, and my numbers, be incorrect? He believed in what later came to be called 'route one' football and he found evidence to support that belief. But true insight only arises from the search for disconfirming evidence – why might the long ball be the *wrong* way to play? Reep wanted to see football as comparable to mechanized production, to see the pitch as a factory where producing more with less was the main goal and where profits depended on maximum efficiency. And he soon began collaborating with managers who thought just like him.

This was where Reep differed from another outsider who attempted to analyse his sport: Bill James, the baseball statistician whose work – as made famous in the film *Moneyball* – went on to influence Billy Beane, the Oakland A's, the Boston Red Sox and the entire game of baseball. For James, the point was to take the numbers and find out what truth they contained, what patterns emerged, what information could be extracted that might change the way we think about the game.

Reep's quest to use the numbers to inform strategy fell short because he was an absolutist, determined to use his data to prove his beliefs. He needed to abandon his idea that he was looking for the one general rule, a winning formula, and learn to seek the multiple truths and the falsehoods in the numbers themselves.

Reforming Beliefs with Data and Analysis

We too are products of our time. We live in an age of Big Data, where our entire medical histories can be put on a memory stick, our musical tastes and photo albums exist in the ether of cloud computing, advertisers have access to our interests and hobbies through social media and supermarkets know our every shopping habit. Analytics are now a critical part of business in countless industries, from medicine to manufacturing and pharmacy to retail. Football is trying to manage the datafication of life in the twenty-first century.

We can push and explore the data further than Reep and Benjamin ever did. To cast the sport in a brighter, truer light, the numbers game requires more than simply counting the actions that happen on a football pitch, but rather looking for patterns in the data from huge samples of information. It also means accepting that certain elements of football are contingent and, if necessary, applying sophisticated statistical models with the help of advanced software and powerful computers.

The aim of the analytics game, though, has changed. If Reep wanted to help teams overcome football's inherent inefficiency based on what he already believed about the game, his heirs want to employ information – pure, cold fact – to establish whether what we think we know about football is actually true. Analytics is not about trying to use the numbers to prove a theory, but to see what the numbers actually tell us, to discover if our beliefs are correct, and if they aren't, to inform us what we should believe instead. As with any journey of discovery, challenging accepted wisdom can be unsettling.

Take the 'fact' that teams are most vulnerable just after they have scored. It's a statement found in football all over the world, and one born of the tricks played by our minds.

The human brain is an analytical modelling machine of the type developed by betting companies. We all naturally build databases and save them on the hard drives between our ears, and then we use them to come to conclusions based on evidence. But, as a forecast and rule producer, our inbuilt computer has its flaws. Our brains are wired to remember and overvalue those events that are most startling and vivid. Events that actually happened are more easily recalled than those that *could* have happened. Our personal theories and views are naturally confirmed: we do not believe it when we see it, rather we see it only when we believe it.

It is here that the numbers come in.

Think of all the football games you have seen: in the overwhelming majority of cases, when a team has gone ahead it has not surrendered the lead immediately. Sometimes it happens, and in spectacular style. Take the game between Bayer Leverkusen and Schalke 04 in April 2004: Hans-Jörg Butt, the Leverkusen goalkeeper and, curiously, regular penalty taker, had just put his side 3–1 up from the spot. He jogged back to his own area, high-fiving all of his teammates, milking the adulation of the crowd. Mike Hanke, the Schalke striker, was unimpressed. He waited for the whistle and, with Butt still sauntering back into position, shot straight from kick-off. Suddenly, it was 3–2. Teams are always most vulnerable when they have just scored, see?

Scholars Peter Ayton and Anna Braennberg of City University London disagree. They analysed 127 Premier League games that ended in a 1–1 draw, and logged when the opening goal and the equalizer were scored. They divided the remainder of the game, after the first goal, into quarters. So if a team took the lead in the tenth minute, the rest of the fixture would have four twenty-minute quarters.[10] According to cliché, the majority of equalizing strikes should have come in the first quarter. The numbers, though, show that the reverse is true. It is immediately after they have scored that teams are *least* likely to concede.

The idea that a team is most vulnerable after it has scored, though, is just one of the myriad myths that pervade football, pearls of received wisdom that are accepted unquestioningly as truth. No doubt José Mourinho, ever the iconoclast, would

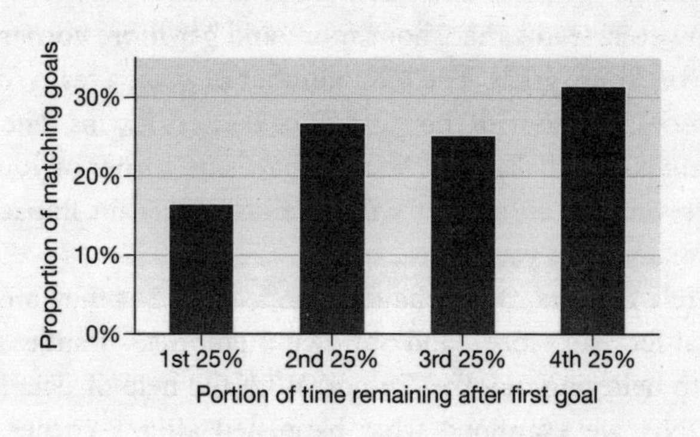

Figure 2 Do scoring teams concede immediately?

put the value of corners into that category. For a manager whose teams have, occasionally, seemed a little over-reliant on set pieces – particularly given the cost at which they tend to be assembled – the Portuguese has seemed to us to be a little scornful of the passion with which corners were greeted in his temporary homeland. 'How many countries can you think of where a corner kick is treated with the same applause as a goal?' Mourinho memorably asked. 'One. It only happens in England.'[11]

He is quite right: corners, in the Premier League and the Football League, are seen as almost the next best thing to a goal. They are greeted by supporters with a throaty roar, their enthusiasm evident, their belief that, at last, the breakthrough is imminent. And why not? After all, just watching the procession of goals from such set pieces on *Match of the Day* is proof that they are reliably profitable. Aren't they?

Well, no, they aren't, as it turns out. The data do prove that corners and shots on goal go hand-in-hand – a team that shoots more will have more corners, and vice versa – as our graph of ten seasons' worth of Premier League matches shows.

However, teams that shoot more and get more corners do not score more goals. The total number of goals a team scores does not increase with the number of corners it wins. The correlation is essentially zero. You can have one corner or you can have seventeen corners: it will have no significant impact on how many goals you score.

Surely corners cannot be that ineffective? But they are, for all that football's lore – and our own memories – want to trick us into believing that they're not. With the help of data from StatDNA we examined what happened after a corner was taken in a sample of 134 Premier League matches from the 2010/11 season – a total of 1,434 corners.[12] We were expecting

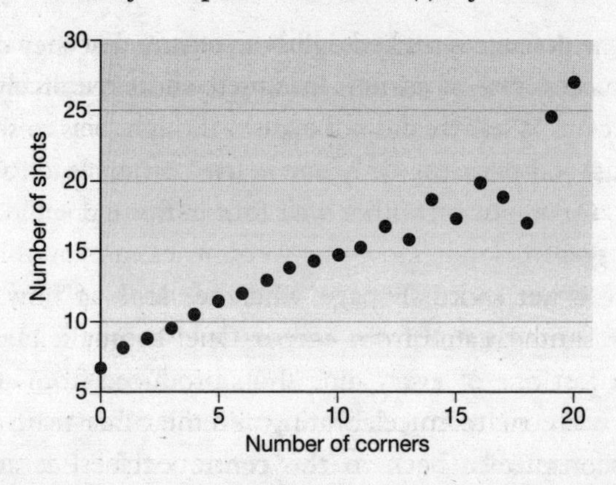

Figure 3 The connection between corners and shots, Premier League, 2001/02–2010/11

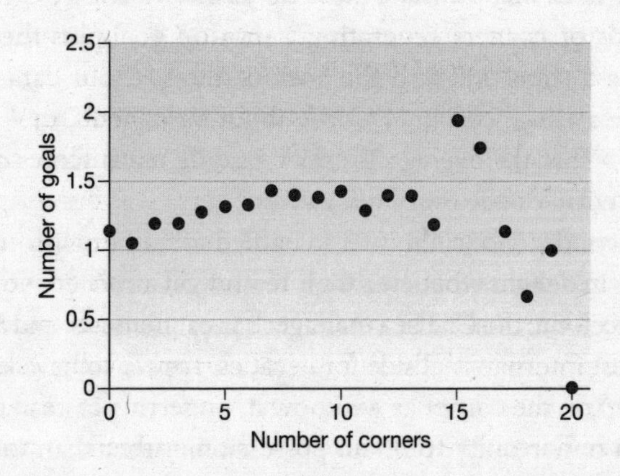

Figure 4 The connection between corners and goals, Premier League, 2001/02–2010/11

to see proof of the following: corners lead to shots, shots lead to goals. Corners, then, should lead to goals.

We expected a degree of slippage. Not every corner leads to

a shot: the defence is packed tightly to ensure that they do not. So the success rate of corners leading to shots is unlikely to be 100 per cent. What we did not expect, though, was to see that it was just 20.5 per cent. Only one in five corners lead to a shot on goal. Or, to put it another way, four in five did *not* lead to a shot on goal.[13]

There is yet more slippage when we look at how many of these shots created from corners lead to goals. Here, we see that just one of every nine shots produced from corners end up with one team celebrating and the other team trudging, disconsolately, back to the centre circle. Put another way: 89 per cent of shots on goal produced from corners are wasted.

How does that translate into real terms? When we combine the odds of corners generating a shot on goal plus the odds that these shots will find the back of the net, our data show that the average corner is worth about 0.022 goals, or – more simply – that the average Premier League team scores a goal from a corner once every ten games.

No wonder Mourinho was so baffled to find English crowds roaring in delight whenever their team won a corner. No wonder Barcelona, the Chelsea manager's great nemesis, and Spain, the finest international side for decades, appear to have largely given up on the corner as we know it, preferring instead to use it as an opportunity to retain possession, rather than to hoist the ball into the box. Corners are next to worthless; given the risk of being caught on the counter-attack, with your central defenders marooned in the opposition's box, their value in terms of net goal difference is close to zero.

Next time your side wins a corner, think twice before urging your tallest players forward. It may be better to play it short, to retain possession, than to hit and hope. The numbers can help us see the game in a different light. What we have always done is not necessarily what we must always do.

What Lies Ahead

That is just a snapshot of what football analytics can do; parlour tricks compared to the deeper findings the numbers are capable of delivering. The football science that started decades ago is expanding, exploring all the time. Where the Wing Commander thought that he could use his system to unearth the perfect way of playing, to rationalize the chaos of the game, his successors – the men (and women) who gather in Boston every year, who study the endless supply of data from Prozone and Opta – believe they can use their information and knowledge to play the game better, to challenge its myths, to see it more clearly.

Far from being a game that cannot be analysed, that is too fluid and too complicated for the numbers to be of any use, football is ripe for dissection, on and off the pitch. Some clubs understand that, companies like Opta and Prozone understand that. Money is pouring into analytics, and all those millions and millions of data points are the reward.

A storm is gathering in football. It is one that will wash away old certainties and change the game we know and love. It will be a game we view more analytically, more scientifically, where we do not accept what we have always been taught, but where we always ask why. The game will look the same, but the way we think about it will be almost unrecognizable.

Professional sports leagues have been slow to catch up with mainstream society in terms of using Big Data to draw big conclusions, and football still lags behind baseball, for example.

Clubs are drenched with a torrent of information, struggling to work out what it can teach them, and what it all means.

There is no secret recipe for success locked in the numbers. There is no winning formula. There is no right answer to football. But there is a way of making sure we are asking the right questions.

Consider this book a manifesto for football's future, a road map for what is to come, a guide not to what the numbers say, but to what we can make the numbers do. All that money has been spent gathering information. Now it is time to sort it, to assess it and to analyse it. To find out what it says.

And it says an awful lot.

It can tell teams whether to shoot more or shoot less, clubs whether to sack their manager or keep the faith, chairmen whether that multimillion-pound striker is really worth the money and the hassle. These are questions that have been asked throughout the history of the game, and tradition and faith provided the responses. It is only now, though, that we have not only the numbers but also the techniques to generate answers.

This is an early dispatch from the front line of this reformation.

We offer a glimpse of what the future might look like, of what these new truths might be. We will look at the work of a host of eminent scientists and academics who have broken football down into its constituent parts and rebuilt it, and we will offer you the results of our own ground-breaking research into the game we love. We will, we expect, challenge some of your assumptions, but we will doubtless support others. We can answer some questions; others we have to leave open to discussion and debate.

We have come a long way from Charles Reep. Football has

always been a numbers game; the Wing Commander was right about that. Much of what we see can be counted; much, though not all, of what can be counted does, as Einstein would have it, count. Now, though, we are starting to know why and how it counts.

Welcome to the reformation. We'll help you keep score.

Before the Match: The Logic of Football Numbers

I.

Riding Your Luck

Toeval is logisch. [Coincidence is logical.]

Johan Cruyff

In the relative anonymity of the seventh tier of Italian football, Loris Angeli, goalkeeper for US Dro, prepares to face the fourth penalty of a heart-stopping shootout. Michael Palma steps up to take the kick for Termeno, Dro's opponents. If he misses, Dro will be promoted.

He takes the kick. Angeli dives to his right, twisting as he does. The ball sails high, towards the centre of the goal. Angeli looks on, helplessly. Palma's shot, though, is a little too strong, a little too high. It knocks against the top of the crossbar and rockets into the sky. Bereft, Palma crumples to his knees and flings himself to the ground.

The ball reaches the top of its arc, and begins to descend. Angeli lifts himself off his back and almost sinks into prayer, giving thanks for his good fortune. He rises to his feet and flies towards the stand to celebrate the miracle.

The ball lands on the edge of the six-yard box. Palma covers his head in despair.

The ball bounces and spins sharply backwards, towards the

goal. Angeli, delirious and oblivious, faces Dro's supporters and pumps his fists as he celebrates.

One bounce, and another, and the ball rolls inexorably towards, then over, the line. Palma peeks, turns, checks with the referee. The goal, a ridiculous, improbable one, is given. Dro miss their next kick. Termeno are promoted.

Football truly is the coincidental game. As we will see later in the book, goals are rare and precious events, ones that clubs spend millions attempting to guarantee. But they are also random. They can defy explanation and disregard probability.

That is not only true down among the lower levels of Italian football. It happens all over the world and it happens all the time. There is the case of Adam Czerskas, a little-known Polish striker, who benefited from football's randomness to score from twenty-five yards with his back as he charged down a clearance. Gary Neville and Paul Robinson suffered from it when the Manchester United player's simple back-pass hit a divot in a Zagreb pitch, bounced over the goalkeeper's swinging foot and condemned England to defeat to Croatia and, ultimately, to missing out on Euro 2008.

Every team, every fan, has seen both sides, but Liverpool, a club all too familiar with fate, can provide two of the neater examples from recent years. On 17 October 2009 Rafael Benítez's team were in the opening phases of a Premier League game at Sunderland when Darren Bent took a snap shot from the edge of the area. Glen Johnson, the Liverpool defender, tried to block the effort, but failed. Instead it clipped a large red beach ball that had drifted on to the pitch and into Pepe Reina's box. The deflection wrong-footed the Spanish goalkeeper, and Liverpool were 1–0 down. Benítez's team had fifteen shots that day, compared to thirteen for the home side, and seven corners to one. And yet, they lost – to a goal scored by a beach ball.

Still, Liverpool cannot complain too bitterly. On the other side of the ledger, they also benefited from an equally improbable, once-in-a-lifetime occurrence just four years previously, on one of the happiest nights in the club's history. Benítez's team came back from three goals down to AC Milan in the 2005 Champions League final, scoring three times in six second-half minutes, to produce what became known as the 'Miracle of Istanbul'.

Even an Everton fan will admit that Liverpool's eventual victory that night was outstanding. But whether it truly was miraculous, or simply extraordinary, is a rather different matter.

In seeking to explain quite what happened, most would point to Benítez's decision to introduce Dietmar Hamann at half-time, his tactical shake-up, his stirring dressing-room speech, or perhaps the superhuman determination of Steven Gerrard, the Liverpool captain – his refusal to wilt, his utter denial of the prospect of defeat.

We cannot test these theories, plausible as they may be. There is no way to examine scientifically what would have happened if Liverpool had not introduced Hamann, or if Benítez had said something different, or if Gerrard had given up hope.

Besides, to do so would miss the point. Maybe Liverpool were fortunate that, on the one unforgettable occasion Milan conspired to throw away a three-goal lead, it was against them, just as they were unlucky that it was during their game at the Stadium of Light that the beach ball wafted on to just the right spot on the pitch to fool Pepe Reina. But that they were present does not suggest the favour or fury of some sort of higher power. There is no special explanation. Beach balls and glorious nights in Constantinople are just outliers on the sea of football data. If you play or watch for long enough, the odds are that these things – that *everything* – will happen sooner or later.

Yes, it's unlikely that on a normal day at the office a beach ball will score a goal, or Milan will let a three-goal lead slip in six minutes, or that Robinson will be fooled by a divot or Czerskas will score with his back or Palma's penalty will hit the bar, rocket skyward, and then meekly roll in. But as Cruyff knew deep in his footballing bones, there is a consistency to the randomness that defines the sport. In football, miracles do happen.

Why Einstein Was Wrong (Sometimes)

Scientists might seem an unlikely group to take much of an interest in football, but there exists a shadowy subset of academics with a serious, persistent curiosity about the game. Scholarly research on football has appeared in countless academic journals across a vast array of fields, including economics, physics, operations research, psychology and statistics. And serious scientific research into the game is on a steep upward trajectory.

Depending on their training and tools, the scientists have developed different ways of understanding the role of predictability and randomness in football, but the essential issue many of them are discussing is a shared one. It also happens to be the same one that Charles Reep, our flawed Football Accountant, was trying to tackle: are football matches and league championships decided by skill, or are they decided by luck?

This is one of the key questions for understanding football, if not *the* key question. If the game is more about skill, then the competition has a logic to it: ultimately, the best team will win. If it is more about luck, then what good does it do for an owner to spend millions on players, for a manager to drill them into perfect harmony and for fans to roar themselves hoarse encouraging them to victory?

Most of us would prefer it to be the former, from managers who sell themselves on their ability to shape destiny, to players determined to outstrip their peers and forge their own place in history. For all the delight that fans take in football's anarchic streak – Greece winning the 2004 European Championship, North Korea beating Italy in the 1966 World Cup – the very idea of being a supporter relies on there being a logic to the game: if your team buys the best players and hires a great manager, the trophies will follow.

In our attempts to discover how great a role chance plays in football, we came to a very different answer, however. We have visited betting parlours and laboratories, and we have encountered many of those scientists who share a passion for the beautiful game. We have examined tens of thousands of European league and cup games over the course of a hundred years and World Cup matches played by dozens of countries since 1938. And we have come to the conclusion that football is basically a 50/50 game. Half of it is luck, and half of it skill.

That is the sort of finding that makes all humans – not just football fans – uncomfortable. Even Albert Einstein, when confronted with the randomness of quantum mechanics, had a hard time believing in chance. 'I, at any rate, am convinced that He does not throw dice,' he famously wrote.

If even Einstein found uncertainty unsettling, no wonder football fans find it hard to accept, preferring instead to concentrate on something soothing – and, crucially, explicable – like beauty.

Football is a sport obsessed with and distracted by beauty. Most fans would – or at least *say* they would – prefer to see their team lose well than win poorly, the sort of attitude of which the great American sports writer Grantland Rice would have approved when he noted, 'When the One Great Scorer comes to mark beside your name/He marks – not that you won or lost – but how you played the game.'

Those teams who are adjudged to capture football's beauty are revered, regardless of results – the Magic Magyars of 1954, the Total Football Holland produced in the 1970s, Brazil of 1970 and 1982, modern-day Barcelona – while others, like Greece in 2004, Italy and West Germany of the 1990s and even Stoke, are reviled for their dour, pragmatic view of the world.

The problem is that beauty is a diversion, and one that can obfuscate the facts. Take the 2010 World Cup final, a game in which Holland produced a display of such flagrant brutality that even Johan Cruyff, that logician of coincidence, was moved to sneer. It was, he said, 'ugly, vulgar, hard, hermetic, hardly eye-catching . . . anti-football'.[1] The high priest of Total Football was clearly ready to excommunicate Nigel de Jong and John Heitinga.

But Cruyff's analysis misses the point: Holland's approach in Johannesburg would have paid off in spectacular fashion had Arjen Robben not fluffed a chance to give Bert van Marwijk's side the lead in the eighty-second minute. The beasts would have done what the beauties never could, and taken the World Cup back to the Netherlands. It may not make for good viewing, but ugliness is no barrier to success. To paraphrase Reiner Calmund, the bombastic former Sporting Director at Bayer Leverkusen, football isn't figure skating. There are no points for style.

Beauty can be a *by-product* of successful teams, but just as it is not sufficient for winning games, neither is it necessary.

We cannot analyse beauty – it is subjective – but we can analyse playing effectively, presuming by 'effectively' we mean things like winning the ball and retaining it, earning free kicks, taking shots and eventually scoring goals. Yet even then we find that, often, doing most things right on the pitch is not enough to win a match.

The instances of teams finding themselves comfortably on top of a game but somehow contriving to lose are numerous. Chelsea managed to lose at Birmingham in a 2010 Premier League game after taking twenty-five shots on goal to their opponents' one. That one was the only one that went in. In Germany a year previously Hertha Berlin mustered seventeen attempts on Köln's goal, compared to just two for their opponents, and still lost. On April Fools' Day in 2006 Zaragoza registered an impressive twenty-nine efforts against Villarreal, and lost 1–0. Football is full of examples of the 'wrong' team winning – the United States beating England in the 1950 World Cup, Cameroon overcoming Argentina in 1990, Wimbledon shocking Liverpool in the 1988 FA Cup final.

Or, more recently, Chelsea lifting their first ever Champions League title after finding themselves defending for 180 minutes against Barcelona in the semi-final and then 120 against Bayern Munich – in Munich – in the final. Against Lionel Messi, Xavi Hernández and Andrés Iniesta, Chelsea at times ceded almost 80 per cent of possession. Across both legs, Barcelona hit the woodwork five times, missed one penalty and a host of chances. Against Bayern, Chelsea found themselves under siege once again, and survived.

The venerable German newspaper *Die Zeit* described Chelsea's win as 'undeserved; more than that, a farce'. Their victory would, they said, 'enter the history books as a football accident'. That night at the Allianz Arena, Bayern recorded thirty-five shots to nine, and had twenty corners to Chelsea's one. Chelsea scored from theirs, obviously. 'Football is just not fair,' said Wolfgang Niersbach, President of the DFB, the German football federation.[2]

That is the thing with football: it does not always reward those who take more shots or complete more passes. It only repays those who score goals. As the *Guardian*'s Richard Williams wrote after that night in Munich: 'Football is a contest of goals, not aesthetics. We love it when the two elements are combined, but that is not the primary purpose of the exercise.'[3]

These examples are one-offs, coincidences, just like the beach ball and the miracles and the missed penalty that ended up as a goal. But this is how we – and those scientists with that unlikely interest in football – choose to react when we encounter randomness: we do not ignore it, or attempt to explain it away as the work of the gods, or concentrate on beauty instead. No, we gather up the coincidences into a large enough set and apply analytical tools to try to understand them. And when we do, we find that – just as Cruyff said – there is a logic to coincidence.

This takes two forms. It applies at the level of leagues and seasons, across cup competitions, where the distribution of goals is reliable and incredibly predictable; and, of more concern to most fans, it applies to individual games, to home and away ties, where the role of chance in producing goals is considerable. In fact it's about 50/50. Half of the goals you see, half of the results you experience, are down not to skill and ability but to random chance and luck.

There are two routes to success in football, we have found. One is being good. The other is being lucky. You need both to win a championship. But you only need one to win a game. The correspondent from *Die Zeit* was right: the history of football is a record of football accidents that follow Cruyff's dictum. Toeval is logisch.

Why Footballers Are Like Prussian Horses

To explain how coincidence and chance allow us to predict what might happen over the course of a league season, we have to take an unlikely detour: to a Prussian military cavalry yard at the tail end of the nineteenth century and the mind of a Russian economist, via the theories of a French mathematician.

Just like professional footballers, cavalry horses lash out occasionally. When they do the consequences can be rather more serious than the injuries sustained in a clash on a football pitch, as the Prussian army discovered in the twenty years from 1875. In that period 196 soldiers met their demise at the hooves of their trusty steeds. These should be purely arbitrary events: military personnel should be experienced enough around horses to recognize when the steeds were spooked, nervous or under fire, and at no point did the army establish that its soldiers were systematically at fault and responsible for their own deaths. No, each of them was accidental and senseless, just an unlucky Prussian standing in precisely the wrong place at the wrong time. No pattern; just coincidence.

It was a Russian political economist of Polish heritage, Ladislaus von Bortkiewicz, who gathered the horse-kick data at the close of the nineteenth century to take another look at the apparently random pattern of deaths.[4] He created a famous data table with 280 boxes (14 corps × 20 years) that showed the number of annual deaths in each corps. When he looked at the boxes, he saw very quickly that the majority of the boxes (51 per cent) were empty, indicating that there were no deaths in that unit that year. A little less than a third had one death, 11 per cent had two, 4 per cent three deaths, two boxes had four, and none had five or more fatalities.

After looking at the table long enough, Bortkiewicz conjectured that there was a logic to the seemingly random coincidences, and that there was a consistency to the randomness. The Russian's insight was to employ a probability equation developed by a French mathematician, Siméon-Denis Poisson. In his *Recherches sur la probabilité des jugements en matière criminelle et en matière civile* (Research on the Probability of Judgments in Criminal and Civil Matters), Poisson sought to describe mathematically the number of matches that would occur as the top cards of two shuffled decks of cards were turned over pair by pair fifty-two times.[5]

Using his cavalry data, Bortkiewicz stumbled upon something the Frenchman had not: that Poisson's equation could give rise to a law of small numbers – a prediction for how many times any rare event could be expected to happen in a given time or place. We can predict the overall frequency and the distribution of random events – how often they occur and how likely they are to do so – as long as we are trying to analyse an event that happens infrequently but also consistently and independently enough to establish a base rate.[6]

Horse kicking is one such event. In Bortkiewicz's data, death at the hooves of the Prussian steeds happened at a rate of around 0.70 per corps per year. Combining that with the Poisson distribution, Bortkiewicz found a remarkable match between the actual distribution of deaths and the predicted distribution. In other words, Poisson's equation provides us with a way to forecast rare and uncertain events.

What does this mean? It means that what appeared senseless, random, is actually subject to a predictable pattern. Bortkiewicz knew *nothing* about the condition of the hay, grass and feed, the amount of exercise and training, the horses' measurements or breeding, or any of the things you think might make

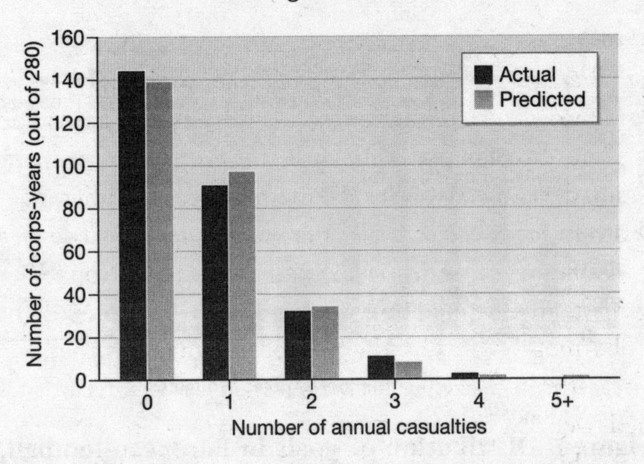

Figure 5 Distribution of fatal Prussian cavalry horse kicks

a difference. All he had was a base rate – how many deaths from kickings occurred per year. Although we can't predict exactly *when* a horse kicking will occur, we can forecast *the overall numbers* exceptionally well. The rare and the uncertain is completely predictable; we know exactly how many of them will happen. Coincidence is logical, just as Cruyff said.

Statisticians have applied the Poisson distribution to many rare events – V-2 strikes on London during World War II; the frequency of traffic accidents; radioactive decay; and so on.

And what does any of this mean for football? Well, just like horse kicks, German bombs and the rate of radioactive decay, goals are rare – quite how rare, we will discuss later – but consistent and independent. Each of them is, at first glance, random. Individually, they are unpredictable. That's what makes them so exciting.

But, by taking the average number of goals per game – 2.66 for the top flights in England, Germany, Spain, Italy and France

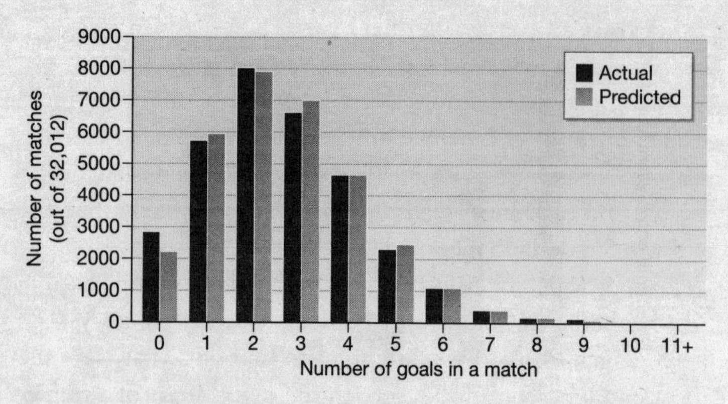

Figure 6 Distribution of goals in European football, 1993–2011

between 1993 and 2011 – and applying the Poisson distribution, we can predict how many games over the last seventeen years saw no goals, how many saw one, how many two and so on. We do not need to know anything about formations, tactics, line-ups, injuries, the manager, or the crowd – none of it – to find that there is a structure to goal scoring. Football might be random, but it is also predictable.

This predictability means that, in next season's Premier League, we know that around thirty games will end goalless, seventy will be won by the only goal of the game, ninety-five will have two goals in total, eighty will see three, fifty-five will have four and fifty really exciting matches will have five or more goals.

How do we know? Well, there are 380 games, and teams score around 1,000 goals a season. Thanks to those kicking horses, the French mathematician and the Russian economist that is all we need to know to extract logic from coincidence.

The Poisson distribution can also apply to individual scorelines.

Let's take an average Premier League Saturday. On 7 November 2010, these were the scorelines: 2–2, 2–1, 2–2, 4–2, 1–1, 2–1, 2–0. Nothing out of the ordinary, but how common are these results once we compare them to many Saturdays for many seasons across a few leagues? Are the 2–1 wins Manchester United and Blackburn recorded that day more likely than the 2–0 win Sunderland recorded over Stoke?

Data provided by Infostrada, a sports media group based in the Netherlands, allow us to calculate the frequency – in percentages – of various outcomes to find out what were the most and least common scores in the ten seasons of Premier League play between 2001 and 2011.

Table 1 Match results by percentage, Premier League, 2001/02–2010/11

	Away goals							
Home goals	0	1	2	3	4	5	6	**Total***
0	8.34	7.58	4.50	1.76	1.00	0.26	0.11	23.55
1	10.92	11.63	5.74	2.66	0.84	0.11	0.08	31.97
2	8.68	9.37	5.03	1.58	0.34	0.08	0.05	25.13
3	4.32	4.37	2.24	0.76	0.21	0.05	–	11.95
4	1.89	1.55	0.74	0.53	0.24	0.03	–	4.97
5	0.55	0.63	0.24	0.16	–	–	–	1.58
6	0.24	0.16	0.11	–	0.03	–	–	0.53
7	0.08	0.11	0.03	–	0.05	–	–	0.24
8	0.03	0.03	–	–	–	–	–	0.05
9	–	0.03	–	–	–	–	–	0.03
Total*	35.05	35.45	18.61	7.45	2.68	0.53	0.24	100

Note: *Rows and columns may not sum precisely owing to rounding.

The most common score is a 1–1 draw – occurring 11.63 per cent of the time – just narrowly ahead of a 1–0, 2–1, and 2–0 home win, a goalless draw, and a 1–0 away win.

Goals really are rare and precious events: more than 30 per cent of matches end with one goal or none. A little less than half of all games end in the home side scoring once or twice and winning; then there is a group of mixed home and away wins and moderately high-scoring draws (1–2, 3–1, 2–2) which occur about 5 per cent of the time each. Finally, there's everything else. On our selected weekend, only one result was truly unusual – Bolton's 4–2 win against Spurs.

This spread of results in the English Premier League, as Figures 7–10 show (the size of the football is proportional to the number of matches), is not markedly different from that

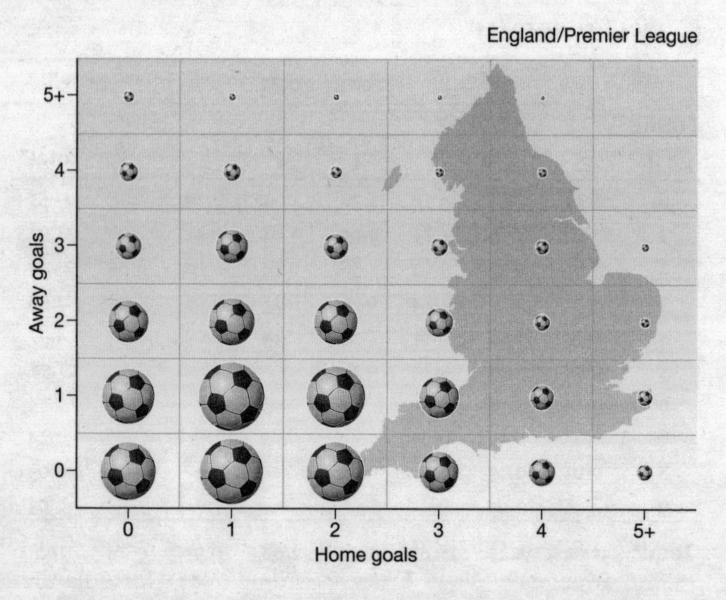

Figure 7 Most common scores in the Premier League

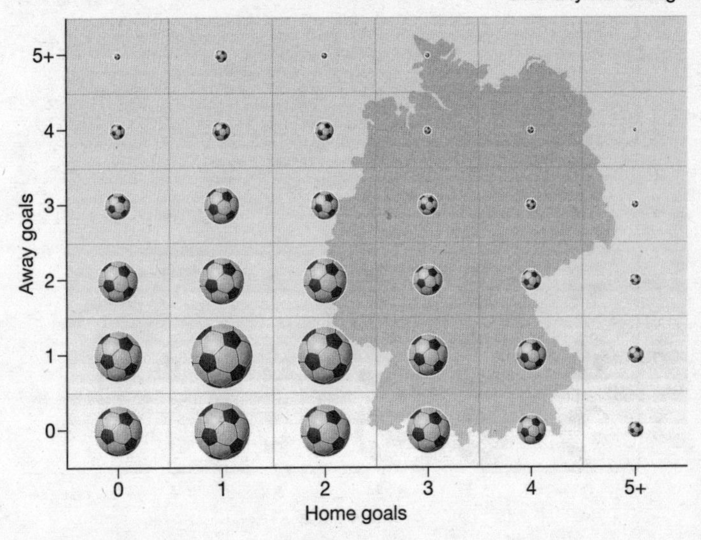

Figure 8 Most common scores in the Bundesliga

found across the top continental leagues during the past decade. That may seem odd. Isn't the football played in Spain different from that played in England? Isn't the footwork of the Spaniards and South Americans plying their trade in Europe's south markedly different from that of the danceless Saxons, Celts and Scandinavians further north? And yet, if you compare the results of the four major European leagues on any given weekend, they show no noticeable difference.

This might surprise football aficionados, but not football scientists. All these results very closely mirror the Poisson distribution. Lots of outcomes of games are possible, but not all scores are equally likely. True, according to the formula there should be 7.7 per cent of games ending goalless – not 8.34 per

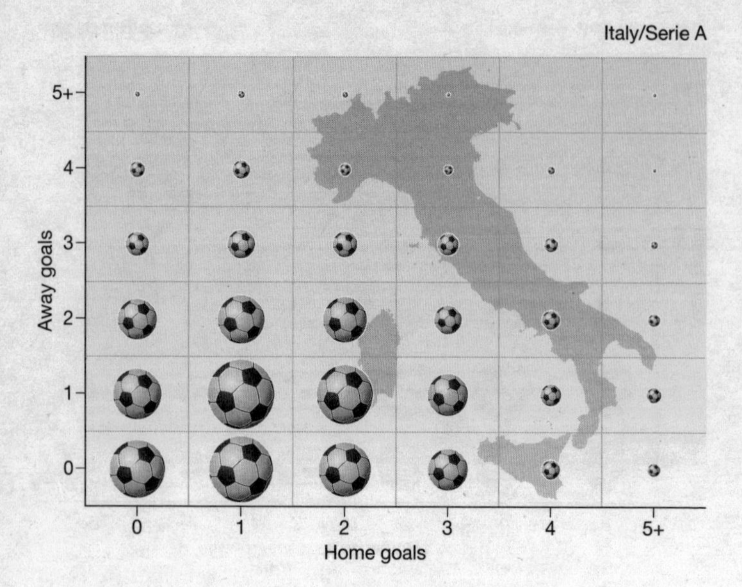

Figure 9 Most common scores in Serie A

cent, as in the Premier League – and 19.7 per cent, not 18.5 per cent, should end with just one goal. But it's pretty close.

That the fit is better for horse kicks than human kicks can be attributed to the importance of drawing matches in football; there are more goalless games and 1–1 ties than the Poisson would expect. There is a slightly greater complexity in the randomness at play in Borussia Dortmund's Westfalenstadion than was present in the bygone Prussian stables. The ball bounces more erratically than the horses bucked.

There is no question that at the level of seasons and leagues there is a mathematical logic to the randomness of goals. That is a fact of football life. That may console managers and encourage gamblers, but what will really concern fans is the other side of the coin: how much of a part will chance play in the game

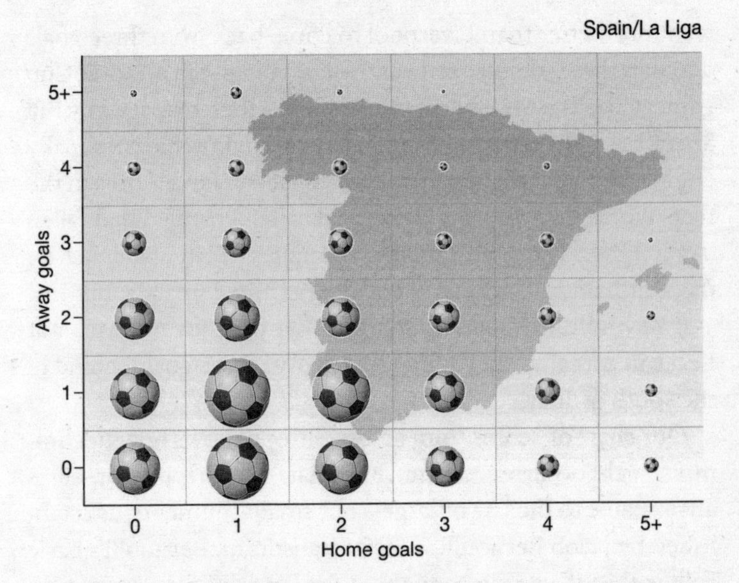

Figure 10 Most common scores in La Liga

you go and see this weekend? Will your team win or lose because of their abilities – or lack of them – or will they simply be betrayed by fate?

What Do Bookies Know?

The 2005 Champions League final against AC Milan was just one of more than 5,000 matches played in Liverpool's history. And yet that was the first time in 112 years of existence the club had recovered from three goals down. No wonder fans afford sanctified status to the Miracle of Istanbul.

Such results are rare, wonderful, but they are hardly unprecedented and they're certainly not miraculous. In 1954, Austria went one better than Liverpool to come back from three goals down in three minutes to beat Switzerland 7–5 in a World Cup game; Charlton once beat Huddersfield – then managed by Bill Shankly – 7–6 after being four goals behind. Eusébio personally engineered a Portuguese comeback against North Korea in the 1966 World Cup, scoring three goals after Portugal had fallen 0–3 behind. There are endless examples: Tottenham led by three at half-time against Manchester United in 2000, but lost 5–3; Kevin-Prince Boateng scored a hat-trick for AC Milan at Lecce in 2011 after they found themselves three goals behind in the south of Italy.

Our chart of scores from across Europe shows how uncommon such occurrences are, but that they happen at all is attributable to the law of large – not small – numbers, as established by Jacob Bernoulli, a Swiss statistician. Bernoulli's basic rule is this: if you do something for long enough, every possible outcome will occur.

Take flipping coins: if you were to flip eight coins in a row, the chances of all eight coming up heads seem remote. One head is a 50/50 shot, of course, or odds of 1/1. Eight consecutive heads, though? That's 255/1.

But what if you had flipped eight coins, four times a week, for forty years – except for a fortnight of holiday every year? You would have flipped eight coins 8,000 times. That's 64,000 flips. The odds that you'd have seen eight heads in a row are no longer so remote. In fact, they're really good. Really, really good. So good that, if you were to go to a bookmaker to place a bet that you had seen eight heads at least once in the last four decades, you'd have to wager the entire GNP of the USA to win six cents. It is close to certain that you'd have flipped eight heads in a row.

Why? Because the more you do something, the more likely

you are to see an unlikely outcome at least once. And so if you play football for long enough – as Liverpool have – you will, eventually, come back from three goals down. Or from four behind, as Newcastle did against Arsenal in 2011 and as Arsenal themselves did against Reading in 2012. There is no law other than that of chance preventing you seeing a team going unbeaten for an entire season, or losing their first twelve games, or, even, a beach ball settling a fixture. Over the long haul, everything is likely to happen at least once.

We know that these events are statistical outliers. But how unusual are they? How rare is it that chance intervenes significantly, with enough influence to turn a match, as happened that night in Istanbul?

Chance is a central element of any given football match, and there are people out there whose very existence proves it. Not coaches or strikers or goalkeepers who always seem to get the rub of the green, but bookmakers and professional gamblers, those men and women whose livelihoods depend on understanding who wins and loses.

A bookmaker's career is built on chance. If matches were predictable, nobody would gamble. Instead, while they are not entirely foreseeable, certain factors – form, injuries, that sort of thing – are known ahead of time. That information provides the basis for setting the odds, and, more often than not, making one team the favourite. These odds tell us something about chance and predictability in sport.

The lower the odds, the more unlucky the favourite for any game has to be to lose, and the more their opponents have to rely on luck to win. When two teams are similar in quality, then luck and on-the-day form decide the contest, and the two teams' odds of winning in the eyes of bookmakers will be identical.[7]

With this in mind, we set about examining odds in football and other sports to establish whether bookmakers believe sports are differently susceptible to chance. We had a suspicion that bookmakers would find football to be unique. Is it harder to predict the outcome of a football match than a baseball game, say? To find out, we collected data from about twenty betting exchanges, along with the final scores from the 2010/11 season in the NBA, NFL, Major League Baseball and the handball Bundesliga in Germany, as well as the football top flights of England, France, Spain, Italy and Germany, and then we threw the Champions League in for good measure.[8] Our first question: how often do favourites, across these different countries and disciplines, end up winning any given game?

In football, it's only a slight majority: a little over half. In handball, basketball and American football the favourites win around two-thirds of their games while in baseball it's a solid 60 per cent. Bookmakers, in other words, pick favourites less successfully in football than in any other sport.

That leads to our second question: why should that be? Is football more susceptible to chance, or are the bookies just bad at setting the correct odds for that particular sport? For that we need to establish more than whether the favoured team won; we need to know whether the odds in football are systematically different. Could it be that favourites win less often in football because they are only narrowly favoured, especially compared to those in other sports?

Favourites aren't all created equal; some are favoured by a lot come match day, others by very little. If coin tossing were a sport, no match would have a favourite and one side's odds of winning would always be listed as 1/1, or, to use the form of odds deployed by some betting exchanges, 2.0.[9] In contrast, in sport, if the team with more skill always wins, their odds would

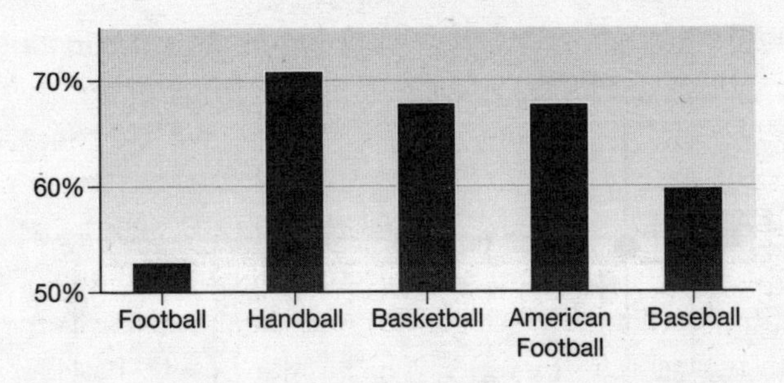

Figure 11 Success rates of pre-game favourites in a variety of sports, 2010/11 season

be listed as 1.0. An even contest, then, will have odds closer to 2.0; if there is an overwhelming favourite, the odds will be closer to 1.0. It is the same with a league, or a sport: those competitions with more certain favourites should have values nearer 1.0; those where the underdogs stand much more of a fighting chance should be further away.

Figure 12 shows the median odds for the favourites during the season for each of the five sports described in Figure 11. The vertical lines show the spread of odds: the bottom of the line is the shortest odds for the biggest favourite of the season; the top of the line is the narrowest favourite for a game in the season.

Football is clearly very different from the other four. Handball has many more large 'overdogs' than football, and the favourites almost always win, with median odds of 1.28; the NBA and NFL have medians of 1.42 and 1.49, respectively. In baseball the spread of odds is most restricted: there are no overwhelming favourites, with the shortest odds being 1.24. But in football, the median odds for a favoured club to win are 1.95.

What does that mean in real terms? Almost half the time in

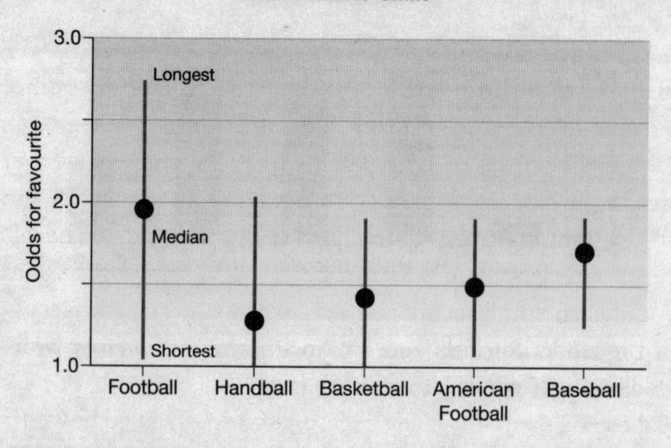

Figure 12 Median and spread of odds across team sports

football, the favourite is not really much of a favourite. Why that should be can be explained by two factors – in football, goals are rare and draws are common. That combination makes setting odds in football much more difficult, and makes favourites less likely to win.

The idea that football's favourites only win about 50 per cent of the time clashes with everything we think we know about the game. Surely, Manchester United against Wigan is not like flipping a coin? Besides, this is hardly a conclusive use of the data: isn't it natural that the bookmakers will get it wrong more often, simply because football, unlike the other sports, has more marginal favourites – teams just fancied to win, but hardly racing certainties?

To find out if that's the case, we need to establish whether strong and weak favourites win at different rates in the various sports. To determine how much a favourite is advantaged over their opponents, we calculate the gap between the odds of the favourite winning and the odds of the underdog winning.

Matches that are toss-ups will have a gap close to zero, while mismatches with hot favourites will have a gap of fifty or more percentage points.[10]

Much like a ratings agency in the financial markets, we went back to the data and separated games into six groups with similar risk ranging from 'blue chips' to 'junk bonds'. The blue chips were games in which a bet on the winning favourite would earn you a secure and very modest return, while a bet on the underdog followed by an improbable victory would yield enough to feed the punter's family for a month. For each of these six slices of a sport's season, we determined how often the differently favoured overdogs won. In other words, we wanted to discover the connection, as in a bond, between risk and performance. The results are shown in Figure 13.

What does this chart show us? Well, football's trend line – representing the relationship between risk and performance for clubs in 2010/11 – sits significantly below the lines of other sports, and it does so regardless of how favoured a team may be.

Take favourites who are fancied 50 per cent more than their opponents: in football, they win 65 per cent of the time, but in basketball, they win more than 80 per cent of games. It is the same across the whole range of risk: the favourite in football is less likely to actually win the match than those in other sports, significantly so in the cases of basketball, baseball and American football where the margin is ten to fifteen percentage points. Football is just a dicier proposition. Bookmakers clearly think football is more susceptible to chance regardless of how lopsided the contest appears to be; and these businessmen know their market.[11]

Our findings take into account only one season, but an even more comprehensive study – undertaken by Eli Ben-Naim, a theoretical physicist at the Los Alamos National Laboratory,

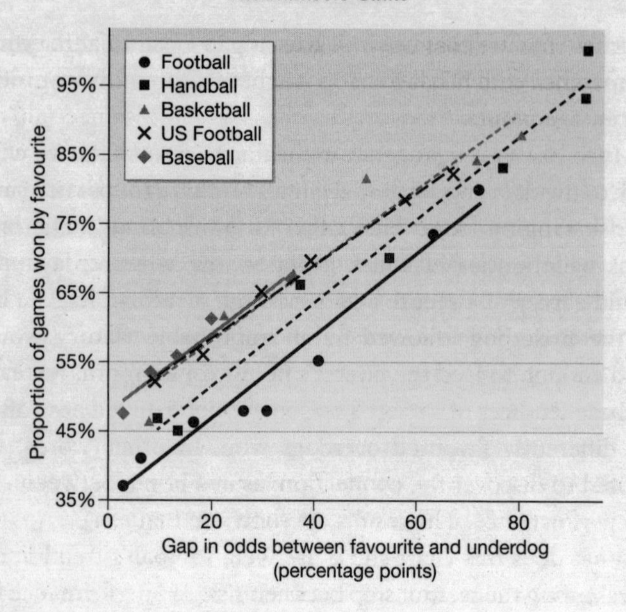

Figure 13 How often do favourites win?

along with Boston University's Sidney Redner and Federico Vazquez – used the entire historical record of a number of sports and came to a very similar conclusion.[12]

Ben-Naim, Redner and Vazquez were interested in how predictable league competitions are, and so their aim was to calculate the likelihood of upsets. As hardcore scientists, they took odds-setting out of the hands of bookmakers and instead constructed artificial sports leagues from computer memory: virtual tables guided by a master equation.

Many virtual seasons of these leagues enabled them to estimate something similar to our gap in odds, based on the records of the favourite and underdog going into the game. They went deep into history, looking at the top flight of English football since 1888, Major League Baseball since 1901, the

National Hockey League since 1917 and the National Football League since 1922. That's 300,000 games in all.

They found, as we did, that football is the most uncertain of the team sports. There are more beach balls and shots against the woodwork in football than in any other game. There are fewer sure things, and fewer no hopers. In more than 43,000 games of football they looked at, the likelihood of the underdog winning was 45.2%. That is the mirror image of our finding.

So almost half the time, the team that is not as well prepared – or has worse players, a raft of injuries, or is just not as good – ends up winning.

On the Trail of Football Scientists

The shadowy subset of scientists with an interest in football have taken things even further than that, in their attempts to determine exactly how much of a role chance plays in any given match.

Take Andreas Heuer, a theoretical chemist at the Universität Münster in Germany, and his collaborators. They noticed the partial mismatch between how the Poisson distribution applied to horse kicks and how it applied to human kicks, and they set out to determine why that should be.[13]

One explanation is that the football data show that teams that have already scored one or two goals become more likely to score a third, a fourth or a fifth – that there is something that happens during the course of the game not captured by Poisson's equation.[14] Take the 2011 Manchester derby – a game City fans will never forget and United's wish they could: did the fourth, fifth and sixth goals United conceded in front of a shell-shocked Old Trafford come from City having what many refer to in football as 'momentum', or was it a fair representation of their greater fitness and ability?

Heuer's team applied mathematical and statistical techniques to twenty years of games from the German Bundesliga as they attempted to discover whether ability and fitness, 'match dynamics' – red cards, injuries, momentum – or what scientists call 'noise', the inexplicable and apparently unpredictable actions of chance, were most significant when it came to understanding goal-scoring patterns. This German team concluded that, mathematically speaking, a football match is a lot like two teams flipping three coins each, where three heads in a row means a goal and 'the number of attempts of both teams is fixed already at the beginning of the match, reflecting their respective fitness in that season'.

In other words, the quality of your squad largely determines the number of shots and a given shot has a one in eight chance of hitting the back of the net, a figure that would look familiar to Charles Reep, our Football Accountant.

Heuer and his team's final results were conclusive. They found that fortune first and foremost, then skill and fitness, then things like momentum ordained whether a team won and by how many goals. That thrashing administered by Roberto Mancini's team to their old rivals was not primarily an expression of their greater ability or an example of how a match can flow in one side's direction. Instead, City were mostly fortunate.

That is a surprising finding to fans who believe a team's skill entirely controls what happens on the pitch, but there is an abundance of other scientific evidence to support it.

A few years ago, two astrophysicists – Gerald Skinner from the University of Maryland in the United States and Guy Freeman from Warwick University – also became interested in match outcomes.[15]

Using some algebra and a sophisticated technique called Bayesian statistics, they set out to determine how often the team with the greater ability actually wins a football match. Or to put it another way, how often the 'wrong' team came away with maximum points. They found – by looking at World Cup games between 1938 and 2006 – that unless a match ends with a three- or four-goal victory, it is very hard to be sure whether the better team won.

Then Skinner and Freeman went one step further. They asked what the probability is that the outcome of a match accurately represents the abilities of the two sides? If results did run in parallel with skill, then what we call an 'intransitive triplet' would be almost unheard of: that is, in a sequence of three games, if Juventus played Roma and won, and Roma played Udinese and won, then Udinese would not subsequently beat Juventus, as we have already established that Juventus are better than Roma and Roma better than Udinese.

Yet Skinner and Freeman found that these intransitive triplets are not nearly as rare as they should be. Partly, they attributed that to the relatively narrow gap in ability – Juventus, Roma and Udinese are separated only by fine margins. It would be different if Juventus were playing Udinese's under-10s, or a local village side. A gulf in ability would make football 'errors', where a bad team beats a good one, much more unlikely.

When Skinner and Freeman looked at their World Cup games, they had 355 triplets of teams playing one another, 147 of which did not involve a draw at any point. Of those 147, 17 were intransitive. That is 12 per cent, which does not sound too much, until you consider that we would expect 25 per cent to be intransitive if the results in all of the matches had been decided completely by chance.

Put in plain English, Skinner and Freeman's data suggested that half of all World Cup matches are decided by chance, not skill. The better team wins only half of the time. Football results resemble a coin toss.

Other scientists have supported this finding. David Spiegelhalter, the Winton Professor of the Public Understanding of Risk at the University of Cambridge, was interested in figuring out whether the final league position in the 2006/07 Premier League season actually reflected teams' 'true' strengths.[16] He wanted to know if the season's champions, Manchester United, really had the best team, and whether the clubs that were relegated (Watford, Charlton Athletic and Sheffield United) really had the three worst teams in the league.

To find an answer, Spiegelhalter had to find out how much of the spread of points in the final league table could be explained by chance alone. Historical records show that 48 per cent of games are home wins, 26 per cent draws and 26 per cent away victories. Spiegelhalter calls this the 48/26/26 law. If we assume teams are indistinguishable by their abilities, we can calculate results for all matches in a season if they were decided according to the 48/26/26 law.

In this imaginary league table, the races for Champions League eligibility and for relegation were more tightly packed than the real table was: proof that there are genuine quality differences among teams. But there is still a certain amount of the spread of the final points that can be explained only by chance. In fact Spiegelhalter's calculations suggest that roughly half the points accumulated can be attributed to fortune.[17]

Of all twenty teams in the Premier League that season, he

found that only Manchester United and Chelsea could confidently be placed in the top half of the table, with odds of 53 per cent and 31 per cent, respectively, that they were the best team. Hardly certainties. At the bottom, he could be 77 per cent sure that Watford were the worst team, but only 30 per cent certain of Sheffield United. That was barely different from Wigan or Fulham, both of whom survived that year. They were no better than Sheffield; they were just luckier.

Meet Professor Luck

No scientist, though, has done more to answer the fans' most crucial question than Martin Lames. A dapper man in his early fifties, with salt-and-pepper hair and a face framed by smart spectacles, Lames is Professor of Training Science and Computer Science in Sport at the Technical University Munich. That may not sound exciting but, in reality, along with his work for FC Augsburg and Bayern Munich, it means he watches football for a living, all in the name of science.

Lames has spent years developing computer and coding systems that allow researchers to record and analyse what happens on the field of play – and why. One of his favourite topics is luck.

Along with a team of collaborators, Lames has used his technology to record instances of good and bad luck on the pitch.[18] Goals in particular are tailor-made for this kind of analysis: some are clearly the result of hard work on the training pitch or superhuman vision from a wonderfully gifted player, and others, well, aren't: an unanticipated deflection, a spilled cross, a missed tackle, a backspinning ball.

To assess how much luck plays a role, Lames and his

match-watching, goal-scrutinizing collaborators defined shooters' good fortune as resulting from one of a total of six goal situations where shooters did mean to score, but the goals they did score had a strong and detectable element of being 'unplanned' or 'uncontrolled'.[19]

Lames and his team have watched videos of over 2,500 goals over the years, coding each one for instances of luck.[20] Alex Rössling, one of his assistants, explains how this process works in practice:

> Everyone saw that the beautiful first goal of the [2006] World Cup by Philipp Lahm bounced from the upright into the goal, which, by itself, was already a bit lucky. But that the ball ended up with the goal scorer because of a bad pass from an opponent provides additional corroboration that it was a goal that wasn't planned or a goal that could be planned. I also really liked the third goal in that match. The cross from Lahm is flicked on by a defender ever so slightly with his head, so that [Miroslav] Klose was able to head the ball because of that failed defensive move; the header then bounced off the goalkeeper and Klose scored on the rebound.[21]

And so, after all those hours and hours of watching goals, how many did Lames and his team qualify as fortunate, as owing more than a little to luck? The answer is 44.4 per cent, though that varied a little from league to league and competition to competition. Lucky goals are particularly common when the score is 0–0. 'That is when teams are still playing according to their system,' says Lames. 'Something coincidental has to happen for a goal to be scored.'[22]

So, about half of all goals contain a detectable, visible portion of good fortune. Football, both goal scoring and favourites

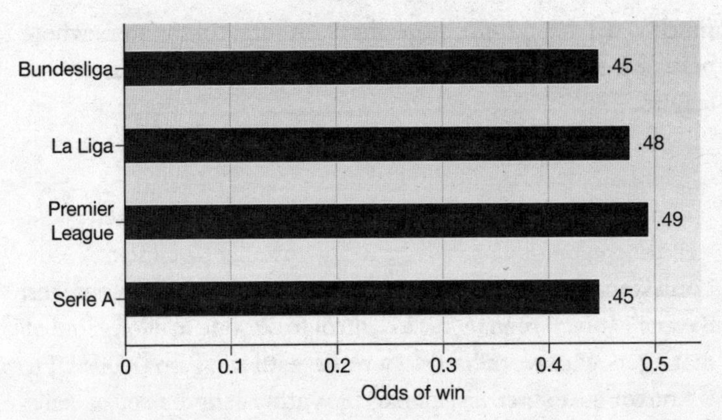

Figure 14 Odds of win for team with more shots in a match, 2005/06–2010/11

winning, is a 50/50 proposition. The game you see this weekend, the one that will leave you in a state of utter jubilation or bitter disbelief, might as well be decided by the flip of a coin.[23]

As the cliché goes, though, surely it is possible to take that randomness and exploit it. The more you shoot, surely the luckier you get?

Not so. We went one step further than Lames to see how often the team that shoots more actually wins a game, by examining data from matches played in the Premier League, La Liga, Serie A and the Bundesliga between 2005 and 2011. That's some 8,232 matches. And what do they show? The team that shoots more actually wins less than half the time. Across our data set as a whole, 47 per cent of teams with more shots won their game. In Italy and Germany, it's as low as 45 per cent.

It is scarcely any better if we limit it to shots on target. Having a greater number of accurate shots than your opponents does make you more likely to win a game, but not by much – the team with more shots on target wins somewhere between 50 and 58 per cent of the time, depending on the league.

Accepting That Football Is Random

Louis van Gaal is the anti-Cruyff. The former Barcelona and Bayern Munich manager is a control freak, one in a long line of managers who work hard to wrestle with the game's odds. He is known as a strict disciplinarian, with a lengthy set of rules about how players ought to behave. Van Gaal believes that football works best when there is absolute and unquestioned discipline on and off the pitch. He even took exception to Luca Toni's table manners at Bayern, when he saw the Italian striker slouched over his plate one lunchtime. 'His back was arched so much, he looked like a question mark,' said one eye-witness. 'Van Gaal saw him and started shouting to sit up. When Toni took no notice, he marched over, grabbed his collar and nearly lifted him out of his seat. Suddenly he was sitting bolt upright. No one said a word. It was incredible.'[24]

Van Gaal sees himself as master of his own destiny. He is not at one with the role fortune plays in football.

Yes, a team needs discipline, and order, and talent, and organization. But there is no denying the role that chance plays in football. It rears its head at the level of leagues and competitions – where the Poisson distribution holds true – and specific matches, where half of all goals can be attributed to luck, and the better team wins only half the time. We have gone from skittish horses to bookmakers and scientists and we have even examined data in a way that has never been done before. The results are in: football is a coin-toss game. Logic and coincidence are evenly split. You've got to find a way to live with randomness in football.

That does not mean there is nothing that can be done. 'What a coach does is attempt to increase the index of probability when it comes to winning a match,' Juanma Lillo, a Spanish manager of philosophical bent, once said. 'As a coach all you can [do] is deny fortune as much of its role as you possibly can.'[25] That means taking your budget, your players, your club, and getting the most out of them that you possibly can. It means spending money wisely, training well, developing tactics and appointing managers in the best possible way.

We cannot control chance. We have to accept that half the time, what happens out there on the pitch is not in our hands. The rest of football, the other 50 per cent, though, is for each team to determine. That is what the billion-dollar industry that surrounds the world's most popular game is built to do. To turn a draw into a victory, to glean as many points as possible, to deny fortune as much as we are able.

We can't all be lucky. But we can all try to be good.

2.

The Goal: Football's Rare Beauty

Everything that rises must converge.

Pierre Teilhard de Chardin

Andrew Lornie was a tinsmith and gas fitter by trade, and a cricketer by inclination. He was not, by any objective measure, a goalkeeper.[1] Still, Lornie, like any good Scot, was not one to turn down a free meal, a drink, and the prospect of a good afternoon's sporting endeavour. And so, when he and his teammates at Aberdeen's Orion Cricket Club received an unexpected invitation to play in the 1885 Scottish Association Football Cup, they leapt at the chance. Alas, the invite had not been meant for them; it was supposed to have been delivered to their neighbours at the Orion Football Club. In the early days of the game, though, such things hardly mattered. The cricketers begged, stole and borrowed whatever kit they could, renamed themselves Bon Accord and on 12 September struck out for Angus – in the middle of a ten-hour rainstorm – to face the might of Arbroath. Lornie would have the unenviable task of keeping goal.

Their opponents, known as the Red Lichties, after the light used to guide fishing boats into harbour from the perils of the North Sea, were an experienced, well-organized side. The faux footballers did not stand a chance.

'The leather,' the *Scottish Athletic Journal* noted, 'was landed between the posts forty-one times, but five of the times were disallowed. Here and there, enthusiasts would be seen, scoring sheet and pencil in hand, taking note of the goals as one would score runs at a cricket match.'

It must have been a disheartening afternoon for Lornie, not least because Arbroath's ground, Gayfield Park, had no nets between the posts: every time the hosts scored, Lornie had to scamper after the ball, retrieve it and bring it back for more. It is testament to his sportsmanship that he kept coming back. His reward was a 36–0 defeat, still the heaviest recorded defeat in British senior football.

Only just, though. As Bon Accord were being hit for six, eighteen miles down the road things were scarcely better for Aberdeen Rovers. They had been drawn against Dundee Harp in the same competition, and they were faring almost as badly. When the match was over, the referee thought Dundee had won by 37 goals to 0; here too, though, the spirit of sportsmanship came up trumps. The Harp players admitted they had only managed a more modest 35. Arbroath would have their place in history.[2]

In one day in 1885, the two sides between them scored seventy-one home goals. A century and a quarter later league football continues to grace both towns. In the season ending in 2011, though, their two teams – Arbroath FC and now Dundee United (the Harp having expired in 1897) – managed sixty-eight home goals between them all season. The weather is much the same, but the goals have dried up in Angus.

The decline in goals is not unique to one corner of Scotland. It is almost unheard of in the modern game to see a team reach double figures; flick through clubs' historical records and their most emphatic victories and heaviest defeats almost always date back several decades. Lornie would not believe it, but goals are rare, and goals are precious, and they are treated as such.

That is why strikers, all over the world, tend to be so revered by supporters and coveted by clubs. Trevor Francis, Britain's first £1 million player, was a forward; so was Alan Shearer, the last Englishman to hold the title of most expensive footballer in the game, after his £15 million transfer from Blackburn to Newcastle in 1996; and Newcastle's Andy Carroll – a striker, naturally – became the most expensive English player when Liverpool bought him for £35 million in January 2011.

Indeed, a glimpse through the list of world record transfers is to read a list of some of the greatest scorers – or providers – of goals in football's long history, from Juan Schiaffino to Diego Maradona and Jean-Pierre Papin to Cristiano Ronaldo.

It is much the same for winners of the prestigious Ballon d'Or, football's most illustrious individual award. Only three even vaguely defensively minded players have been handed the trophy since Franz Beckenbauer won it in 1976 – Lothar Matthäus, Matthias Sammer and Fabio Cannavaro – and all three of them did it in years when they led their countries to victory in a major international tournament. The only goalkeeper ever to win it was Dynamo Moscow's legendary Lev Yashin in 1963. Otherwise, it is a trinket contested by forwards, reward for their wizardry – as in the case of the last-ever winner of the award, Lionel Messi – or their ruthlessness, the trait which helped Andriy Shevchenko, Michael Owen and George Weah claim the trophy.[3]

Football is a sport of chance and fortune, in which all we can hope to do is make the best of what little influence we have. A great striker, though, is seen throughout the game as someone who can seize control of his destiny and his club's, a man capable of taming the randomness. Such a player, like the gems he provides, is rare and precious.

Football's Uniqueness

The goal is more than just football's primary product, the point of all the huffing and puffing over the course of ninety minutes. It's also more than the reason teams buy wonderful, skilful attacking players and managers develop intricate, complex defensive strategies. It is what makes the game what it is. It is something that has to be worked for, that only happens very occasionally, that we spend hours waiting to see.

Football is special, that much is clear. It is not just the beautiful game, but the world's game, a language spoken from the favelas of Rio de Janeiro to the steppes of Asia. We would not have it any other way. But its universal appeal demands investigation and, if possible, explanation. Why is football so enduringly, so ubiquitously, popular? What is it about football that people love?

The answer, of course, lies in the goal. The goal is football. Its rarity is its magic.

Perhaps the easiest way to see what makes football special is by establishing just what it is not. For that, we will need a mechanism to compare it to other, similar sports, defined scientifically as being 'invasion games' that are 'time dependent'.[4] That means, in slightly less convoluted terms, sports that take place on a defined pitch, with a final whistle, and two teams trying to score against each other. Basketball, lacrosse, the two codes of rugby, American football and hockey, both field and ice, are all games belonging to the same genus as football.

But while football is similar to all these sports in broad terms, it is clearly distinct. Football is defined by rare events – goals – but they exist in a sea of hundreds, thousands of extraneous events: tackles, passes, long throw-ins. Football is different because the things that decide who wins and who loses happen only occasionally, while other things – such as passes – happen all the time. And it is this rarity – the lopsidedness between effort and scoring – we believe, that lends football its allure.

But rarity is a subjective concept: if you score once a month and I score once a year, what is rare to you may seem frequent to me. So to establish just how rare goals are, we need to compare football to the other members of its family.

To do so, we collected data on team scores for games over the course of a whole season, in 2010 and 2011, in the top leagues for basketball, ice-hockey, football, American football, rugby union and rugby league. That meant analysing 1,230 NBA games, 1,230 in the NHL, 380 in the Premier League, 256 in the NFL, 132 rugby union matches and 192 from Australia's NRL. We also calculated the ratio of goals (and shots, where possible) per minute, as well as goals per attempt for each sport.

We had to make a few adjustments to make the scoring comparable; American football's scoring system of six points for a touchdown and three points for a field goal or basketball's two points for a basket, three for a long shot and one for a foul shot, for example, had to be transformed so that we could compare them to football scores.

The point was to count the number of times a team scored a goal or the equivalent of a goal. In the simple experiment, we counted up the total number of times a team scored; in the more complicated test, we adjusted these for the relative values of scores. But we needn't have bothered – the conclusions we come to are unaffected by how we do the maths.

Two of the bars in Figure 15 stand out immediately. Basket-

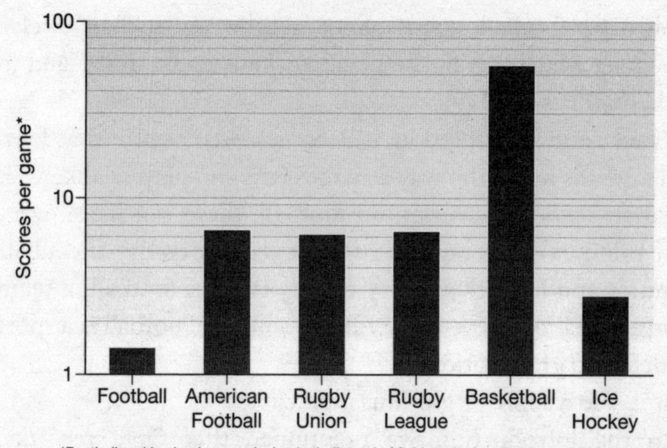

*Football and ice hockey: goals; basketball: total of field goals, free throws and 3 pt shots; American football: total of touchdowns, extra points, field goals, 2pt conversions and safeties; rugby union and rugby league: total of drop goals, tries, conversions and penalty kicks.

Figure 15 Scoring with your team or club

ball is clearly different from the others by a huge margin. If football is a sport of rarity, basketball is a sport of plenty, of frequency, of an almost relentless abundance. There are more scores in basketball than in any other sport by a considerable order of magnitude (note that the left-hand scale in the figure is logarithmic).

But more relevant is the distance by which football anchors the other end of the scale. If basketball's bar looks like LeBron James on a stepladder, football's is Lionel Messi in a pothole, crouching down to tie his boots. It is hardly a dramatic revelation to say that football is the lowest scoring of the team sports. The extent by which it achieves this distinction, though, is stunning.

Just as important is the fact that footballers make fewer attempts to score. In comparison to those other sports where attempted scoring is a relevant statistic, the numbers show that football teams shoot a little more than twelve times per match. In hockey, that rockets up to thirty, and in basketball, 123.

Once time is factored in, it becomes even clearer that football's genius lies in the way it makes fans and players alike wait for their reward. In American football, there is a score every nine minutes on average; in rugby, it is every twelve and a half minutes and in hockey every twenty-two. In football, a team scores a goal once every sixty-nine minutes. Football is a sport of deferred gratification.

It is also a sport of glorious inefficiency.

In the introduction we mentioned that Opta recorded 2,842 events during the 2010 Champions League final between Inter Milan and Bayern Munich. Two of those were goals, both scored by Diego Milito, signed by José Mourinho for more than £20 million the previous summer. Two events of 2,842 that count. That's one goal per 1,421 events. No other sport demands so much effort from a team before anything happens that actually matters.

That is what makes football special, and what makes football what it is. It takes so much effort to score that each goal is celebrated that little bit more joyously, and means that little bit more. That is why the game is so exciting. Any one goal, at any time in the game, can be the difference between victory and defeat, between delight and despair. The goal is football's beauty, and she's a rare and reluctant beauty indeed.

Accounting for the Drought

Thanks to a Basque by the name of Ignacio Palacios-Huerta, we know that, while goals once were plentiful, since Andrew

Lornie's dismal debut, they have been getting ever rarer. What is not immediately clear is why that should be.

Palacios-Huerta is an economist at the prestigious London School of Economics. Some time ago he became concerned with football's main product – goals, and the outcomes of matches.[5] To find out whether there had been any significant changes in how many goals have been scored in the average match since the start of organized football, he did what any good economist would do: he gathered as many numbers as he could and he analysed them. That meant looking at the goals scored in all games in the English professional and amateur leagues between 1888 and 1996. That's 119,787 matches.[6]

Palacios-Huerta focused first on the top tier. His careful analyses of those games showed that goals have declined over the course of football's history. At the end of the 1890s and the start of the 1900s, the scoring rate per match in English football's top flight was plunging, from a high of around four and a half goals per game. It continued to fall until the change of the offside rule in 1925 – reducing from three to two the number of opponents needed to be between a player and the goal line, and therefore making it easier to score – caused the rate to jump up by almost a goal a game. Yet again, that higher rate of scoring eroded towards an average of three goals per match until the outbreak of World War II. When organized football returned at the end of the conflict, there was an increase, but by 1968 the average was once more around three goals a game. By the time Palacios-Huerta's data ran out, it had slipped even further, down to 2.6 goals a game in the 1996 Premier League season.

If that seems obvious, keep in mind that there are strong arguments that would suggest scoring should have gone up over time. Going by other areas of human performance that would be a more than reasonable assertion. Pitches and players are much better maintained than they used to be, equipment is better, and clubs can now cherry-pick the best talent from across the globe. Things generally get better as time wears on.

That's certainly the argument Geoff Colvin makes in his bestselling book about the origins of exceptional human performance, *Talent Is Overrated*: 'Most apparent is the trend of rapidly rising standards in virtually every domain,' Colvin writes. 'To overstate only slightly, people everywhere are doing and making pretty much everything better.' Among the more amusing examples are the fact that 'today's best high school time in the marathon beats the 1908 Olympic gold medallist by more than twenty minutes', or that, in diving, 'the double summersault was almost prohibited as recently as the 1924 Olympics because it was considered too dangerous'. 'Today, it's boring,' he adds.[7]

If Colvin's theory is correct, goals per game should not have gone down. Of course, just as strikers have got better, so have defenders and goalkeepers; but improvements in offensive and defensive performance should have moved in tandem over time, meaning at least as many goals could be expected now as were scored one hundred years ago. That, plainly, is not the case.

Why, then, are goals becoming more and more rare? Rule changes have had only a fleeting effect – the change in the offside law in 1925, the introduction of three points for a win in 1981 and disallowing the back-pass to the goalkeeper in 1992 – if any at all. Likewise, the disruption caused by the two world wars did not alter the long-term trend.

If talent by itself – not tactics or training – had something to do with the increasing rarity of goals, then we should see differences in scoring across the divisions, and those differences in scoring should change over time. The logic goes something like this: let's assume that there was a gap between the aptitude of the players in the top tier and second tier of the Football League around the turn of the twentieth century. Given the nascent professionalism circa 1900, this talent gap was likely to have been quite modest early on. But over time, the rise in salaries, the dramatic growth of resources for training and the global sourcing of players have increased the talent gap between what is now the Premier League and the Championship. Put simply, the difference in the average player's talent between first and second division should be greater now than it was a century ago.

It is reasonable to suppose that a similar trend has taken place since the end of World War II in the difference in ability levels among the second, third and fourth levels of professional football. Logically, then, if skill and talent alone – athletic goalkeepers able to cover more of the goalmouth more quickly, defenders quicker to the ball and more lethal in the tackle, midfielders with more speed and stamina able to track back rapidly and continually – were responsible for the decline in goals, then changes in the relative talent levels across the leagues should mean that their scoring levels should also diverge over the course of the twenty-first century and into the present.[8] Divergent trends in talent should have gone hand-in-hand with divergent trends in goals. So goals should have become *more* rare in the top division than the next division and so on, and differences in goal frequencies should increase over time.

For us to establish whether that is correct, the key assumption – the widening of the talent gap among the tiers of English football – has to be true. And so, for proof, we can look at the FA Cup, the one tournament where the various levels of football skill have met one another for well over one hundred years. Because teams from different divisions regularly play each other in the competition, it also allows us to see if the best really have got better.

Figure 16 shows the number of clubs from the top tier, the second flight and all the lower levels who have reached the FA Cup quarter-finals since 1900. Each trophy represents an average of one club; trophies missing their tops and handles represent proper fractions. Hence, in the first decade of the twentieth century, on average, 4.8 top-tier clubs, 1.7 second-tier clubs and 1.5 non-league clubs made the quarter-finals.

The chart shows that, today, the quarter-final spots and the silverware are being hoarded by the big boys at the expense of their smaller rivals. There are exceptions, of course, such as Millwall and Cardiff City reaching the final in 2004 and 2008 respectively, but the general trend is clear: since the immediate post-war years, the second tier has lost almost a spot and a half to the top division.

This is strong evidence that the gaps in talent and skill among the tiers of English football have indeed widened through the years.

But now the key question: have the growing gaps in talent corresponded with divergences in scoring rates across the tiers of football?

Conducting a series of sophisticated statistical tests, Palacios-Huerta found that, as far as goals go, the top tier and the second were the same. Their historical, year-by-year distributions of goals were identical. Similarly the post-war pattern of goals was shared between all divisions, top to bottom. The general impression was the same, no matter how good players were: occasional bumps, caused by rule changes or the wars, set against a trend towards fewer goals. Skill levels have increased

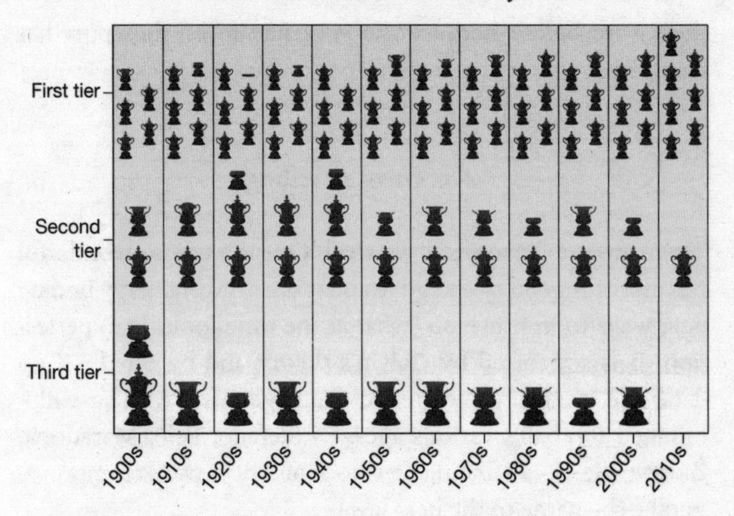

Figure 16 English FA Cup performance by tier, 1900–2012

Note: Based on average number of clubs qualifying for the quarter-finals.

and they have diverged. And yet, the present-day top-tier defender who is so very much better than his 1948 counterpart, and the League Two defender who is only moderately better than his post-war foil, nevertheless thwart goal scoring with equal efficacy. Hence, we can say definitively that the scoring drought in Angus and all other parts of the football world is *not* due solely to the increased skill and athleticism of footballers.

So we know goals, since the end of the Victorian age, have always been rare, and we know that they're getting rarer. We know that is not because of rule changes, major international cataclysms, or the rising level of skill. No, it is something entirely different that is leading to football becoming the most abstinent of sports. Goals are more rare now than they were before because the very nature of the sport has changed.

The Great Levelling

There are two histories of football. One is a tale of wonderful players, of ingenuity and guile and wizardry, constantly finding new ways to improve on (what at the time looks like) perfection. It is supported by Colvin's theory and by our FA Cup data, and it explains the great defining geniuses who have illuminated football's various ages: Di Stefano, Pelé, Maradona, Zidane, Messi – all finding new horizons, new ways to improve, to take the game to the next level.

And there is a second history, one of the men who did all they could to stop them. Not the defenders, but the managers, who dreamed up *catenaccio* and zonal marking and the sweeper system and all the rest: all of it designed to stop the virtuosos showcasing their talents. Even the *tiki-taka* style honed and perfected by Barcelona and adopted by Spain has been labelled a primarily defensive approach – *passenaccio* – because its emphasis is on starving the opposition of the ball.

Players have improved as the game has matured: they run faster, they shoot harder, they dribble quicker and they pass more accurately. And as they have improved, so structures have been built to contain them.

These structures – offside traps, pressing, zonal marking, triangular passing – are the reason that goal scoring has largely withered on the vine. Tactics and strategies have become more complex, cutting off the supply of goals. As individuals have stretched the limits of their own abilities, so teams have found ways to counteract them. As football has developed, it has become a sport in which better and more skilled athletes are more effectively combined, positioned, structured and unified, and as a result, Lornie's heirs have to fetch much less leather out of the back of their nets.

A quick glimpse at typical formations over the years proves as much. There was a time when seven players on any given side were given over to attacking, with two half backs and one full back. That soon morphed into the W-M formation as two attackers were pulled back, and then came the 4–2–4 of Hungary and Brazil, the 4–4–2 so beloved of English managers, and now the trend is to deploy just one striker. Barcelona and Spain do not even do that, since the rise of what has been called the false nine. As the title of Jonathan Wilson's magisterial history of tactics suggests, the pyramid has been inverted.[9]

That says a lot about the nature of the game we love. Where once football was purely an attacking sport, it is now focused on developing a symmetry between scoring *and* not conceding. It has grown into a more balanced game of offence and defence. When tactical changes produced teams that were more defensive but still won (or perhaps won even more), their opponents adapted their playing styles in response. Over time football was discovered as a game that is fundamentally about avoiding mistakes and punishing the other side for theirs.

That bears itself out in the numbers. Had Opta been present at a league game in 1910, we suspect they would have recorded hundreds of touches of the ball from forwards, but very few from a team's ineffectual defenders. A century later, defenders and midfielders see significantly more of the ball then forwards. Opta's figures show that defenders averaged 63 touches of the ball per ninety minutes in the 2010/11 Premier League season, with midfielders on 73 and forwards down to 51.

This is a worrying trend, not least because Palacios-Huerta's findings – together with the switch in emphasis from an attacking to a defensive game – suggest that, at some point, the goal, already threatened with extinction, may die out altogether.

To find out quite how fast that day may be approaching, we decided to update his work (his data ended in 1996). So we gathered more recent information, focusing on football after World War II, and examined the trends in scoring ourselves. Since one season can be unusual for a host of reasons – the weather, luck, a few particularly dire teams – we wanted to be certain we were homing in on a historical trend not distorted by random fluctuations. When we employed a statistical technique known as lowess smoothing that cuts away much of that 'noise', a startling picture emerged.

Instead of the persistent downward trend in goals we have seen over a century and a half of play, in the last sixty years or so there appears to be a levelling off. Goals are not dying. They are plateauing. Scoring has remained essentially stable in the last two decades, perhaps even as far back as the 1970s.

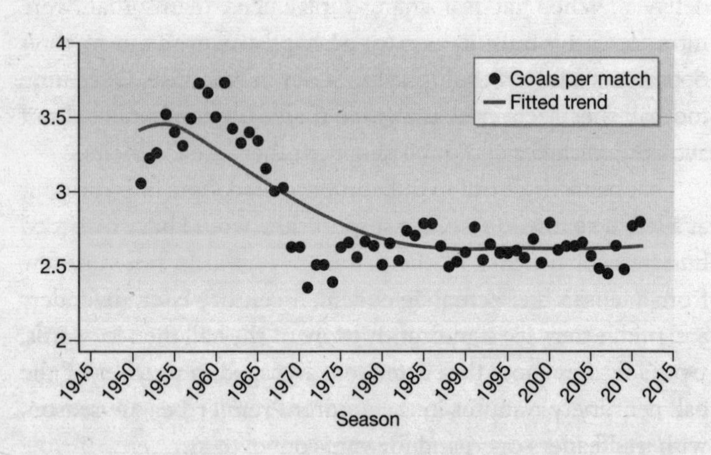

Figure 17 Goals per match, English first tier, 1950–2010

This means a dynamic balancing between two forces: offensive innovation and defensive technology.

Over time, as knowledge about the game spread and successful ideas were copied all over the world, teams have become more alike. Many of the higher scores in the early days of the game had less to do with variations in players' abilities and playing conditions, and more to do with some select clubs having huge advantages in training, setting up tactically, and organizing and coordinating instantly on the pitch. In other words the Orion cricketers were most disadvantaged not by the lack of dribbling and passing expertise, or by the rain and the mud, but by their disorganization and collective tactical ignorance.

Slowly but surely, intentionally and through trial and error – and mostly by eliminating mistakes and weaknesses – teams have become more similar to one another over time.

Looking at the average number of goals scored can be a little misleading, though: a team that scores 0, 0, 0, 6 and 9 goals in five matches has the same average number of goals as a team that scored three in every single outing. The average is interesting, but it does not tell us how many unusual teams and matches there were, and whether the number of these outliers has changed over time.

It has, and considerably. When we calculate the average goal difference in each match in each season of league football since 1888, we see that teams have become more similar in both defensive and offensive output. Teams now win by fewer goals than they used to, with the average goal difference in a match declining from over one to less than half a goal over the last century or so. In one hundred years, the differences between teams have declined by around 50 per cent. If you look at the last thirty years, you'll see that even as the total number of goals has levelled out, goal difference has continued to shrink.

To draw an economic parallel, as their industry has matured, footballers make fewer units of their primary products than they did when their business was young. The trends also suggest that the manufacturing technologies – the best ways of playing – have been diffused over time: through sharing and imitation, along with an opening to a global pool of talent everyone has access to, teams have grown more similar. Football, in this sense, is just another economic sector: today, a Toyota car is scarcely different from a Honda or a Volkswagen; in the very early days of the motor industry every manufacturer used components made to its own specification.

That suggests that one of the sport's great truisms – that the power and wealth of elite clubs has unbalanced leagues across the world – may be a myth, at least when examined from a long-term, historical perspective. If anything league football is more competitive now than it was fifty or a hundred years ago.

Our friends in Arbroath prove this for us: at the top of the football pyramid, the relative rate of improvement for the worst clubs has been greater than that of the best, so there are no longer regular games between fully professional teams and those comprised of tinsmiths, gas fitters and cricketers. Derby, 2007/08 vintage, might have been the worst team in Premier League history, but they were closer in collective ability to champions Manchester United than Birmingham would have been when they propped up the division a century earlier as United secured their first league title.

That increased competition has had one added impact: it has made goals even rarer, even more precious than they were sixty or a hundred years ago. This is one of the great misunderstandings about football: that fans come to see goals. That was what was behind the change in the offside rule, the introduction of three points for a win, or the abolition of the back-pass – a misguided belief that all supporters want to see are goals. What they really want to see are matches in which every goal is essential and potentially decisive.

With the levelling off of total goals and the continued decrease in goal difference, the industry of football has delivered its customers exactly that – tight, low-scoring, nail-biting matches in which no team is guaranteed a thrashing or is facing the insurmountable odds Orion's cricketers-cum-footballers did all those years ago.

Fans may look at the profligate days of the 1890s with longing, assuming more goals equalled more fun. But it is the rarity, the preciousness, of each and every goal that makes them mean so much.

Currently, goals in English football are manufactured at a rate of around 2.66 for every game played, across the divisions and ability levels. Sometimes that goes up a little, sometimes it goes down, but overall it is remarkably stable. So you will see 1,000 goals, give or take, in the Premier League this season, and the season after that, and the season after that. Football seems to have found its equilibrium.

Everything That Rises Must Converge[10]

'I play therefore I am,' the Uruguayan author Eduardo Galeano wrote in his treatise *Soccer in Sun and Shadow*. 'A style of play is a way of being that reveals the unique profile of each community and affirms its right to be different. Tell me how you play and I'll tell you who you are. For many years soccer has been played in different styles, expressions of the personality of each people, and the preservation of that diversity is more necessary today than ever before.'[11]

It is an evocative sentiment, beautifully expressed, but one that is open to misinterpretation. Across the world, there is a powerful belief that foreigners, outsiders, immigrants, often fail to grasp the intricacies and subtleties present in their new league. In England, this credo finds itself crystallized in the 'rainy night at Stoke test'; that is, the belief that certain types of players cannot be viewed as Premier League material until they have shown they can cut it under a downpour at the Britannia Stadium.

This insularity, this semblance of superiority, is not an exclusively English attitude. In Germany, when Frank Arnesen, previously Technical Director at Chelsea, joined Hamburger SV and brought with him Lee Congerton and Steven Houston, formerly scouts at Stamford Bridge, they were accused of not understanding the vicissitudes of the Bundesliga.

Houston and Congerton were intriguing appointments. Houston, a former insurance analyst who trained in sport at the Houston Rockets of the NBA, is one of football's first 'technical' scouts, a man who uses data to assess the opposition, potential recruits and his own players.

We spent some time with them in 2011 to discuss their plans to bring a new brand of analytics to one of Europe's grand old houses, a club so venerable that it is known in Germany as the country's 'dino' – the only member of the Bundesliga present since its inception. It was a difficult season. Things were not going well on the pitch or off it, and the former Chelsea men stood accused of introducing alien, foreign concepts to a league where such an approach was not appropriate. Germany, the Germans said, was different, just as the English think the Premier League is a class apart, and the Spanish and the Italians believe that their type of football is unique.

In some ways, perhaps they are. Perhaps styles change or the frequency with which the referee blows his whistle varies a little. But when it comes to what really matters, they are not unique at all. The strongest leagues in the world, those in Germany, England, Spain and Italy are distinctly similar when it comes to their key traits. In fact, our data show that, despite superficial differences, the most elite leagues are incredibly alike. Everything that rises must converge.

That is not to say that where you are from does not make a difference on the pitch. In 2011 the political economists Edward Miguel, Sebastián Saiegh and Shanker Satyanath examined the connection between civil conflict (political violence) in a player's home country 'and his propensity to behave violently on the pitch, as measured by yellow and red cards the player received'.

Their study's storyline is straightforward: many of today's professional footballers come from poorer countries with significant levels of civil strife and political instability, while others were raised in the rich, stable, democratic countries of the West. Does this affect how they behave when they play the game? The answer appears to be yes. Based on data from the 2004/05 and 2005/06 seasons in five national leagues (England, France, Germany, Italy and Spain) as well as the Champions League, Miguel et al. found a connection between civil conflict in a player's home country and a player's propensity to behave violently on the pitch, as measured by yellow and red cards: as the number of years a country has experienced civil war goes up, so does the average number of yellow cards per player from that country.

'Colombia and Israel are the two sample countries that experienced civil war in every year since 1980, and their players are remarkably violent on the pitch. Inter Milan's Colombian defender Iván Ramiro Córdoba is a case in point: in 2004–2005 and 2005–2006, he collected a stunning twenty-five yellow cards.'

The same pattern appears when the authors look only at players from the non-OECD countries (poorer, less democratic countries in general). While the study doesn't really explain how or why these effects exist, they do provide evidence that players from different countries – with different cultures and political histories – exhibit different behaviour on the pitch.[12]

There is an abundance of data to support that idea. Consider the types of formations typically employed by teams in the Premier League and La Liga. Data from Opta Sports show that Spanish clubs used a 4–2–3–1 formation in 57.8 per cent of all matches they played during the 2010/11 season, while English teams did so in only 9 per cent of theirs.

By contrast, the preferred formation of English clubs was a classic 4–4–2 (used 44.3 per cent of the time). And while the second most preferred formation of Premier League clubs was a 4–5–1 – used in 18 per cent of all matches – La Liga clubs used 4–5–1 in a negligible 1.3 per cent of games. If nothing else, these differences point to contrasting tactical approaches to the game.

Or consider differences in discipline (or, as perhaps an upright Englishman might say, a willingness to dive). When we compare the number of fouls and cautions in England and Spain over the course of the seasons 2005/06–2010/11, we find some noteworthy differences. While the average Premier

League match saw twenty-four fouls, referees in La Liga whistled for a foul thirty-four times per match – a considerable difference of some 40 per cent.

The number of cautions tells a similar tale: while Premier League referees showed 3.2 yellow cards per match over those same five seasons, referees in Spain's La Liga were busy at a rate of 5.1 yellow cards per match – a difference of 59 per cent.

The Miguel, Saiegh and Satyanath study we mentioned above also provides support for these numbers – they find that the incidence of yellow and red cards is systematically higher in Spain than elsewhere, even after they adjust for important factors such as players' positions, age, quality and homeland strife.

But none of these minor discrepancies affect the outcomes of matches. They are incredibly similar across the top level of football in the twenty-first century. The most essential elements of the sport differ very little across countries and leagues. And there is one element more crucial than any other: the rare, the precious goal.

When it comes to goals, though, all those findings, and Galeano's philosophy, do not stand up. It does not matter whether your league has more foreign players or is reliant on homegrown talent; it does not matter if your tactical blueprint was originally inspired by Rinus Michels and Johan Cruyff or by Nereo Rocco and Helenio Herrera, the grandmasters of *catenaccio*; it does not matter a jot if your league is infused with imports from northern Europe and France, like the Premier League, or Brazil and Argentina, like Spain and Italy, or Eastern Europe, like Germany. It may or may not be true that English players are fair, energetic and robust, that Argentines are wily and erratic, that Brazilians are rhythmic and inventive and that South Koreans and Japanese players are hard-working and well organized. None of it is important when we look solely at goals in football's top leagues.

So sure are we that football looks the same among its ultimate elite, we have prepared an experiment on the nature of scoring in the best leagues. To identify the best leagues, we bow to UEFA: their numbers show that, for many years now, there are four leagues that have risen above all others: the Premier League, the Bundesliga, La Liga, and Serie A. Figure 18a shows the goals that were scored in the average match in these leagues in the eleven seasons from 2000/01.

Can you tell which one is which? If not, don't feel too bad.

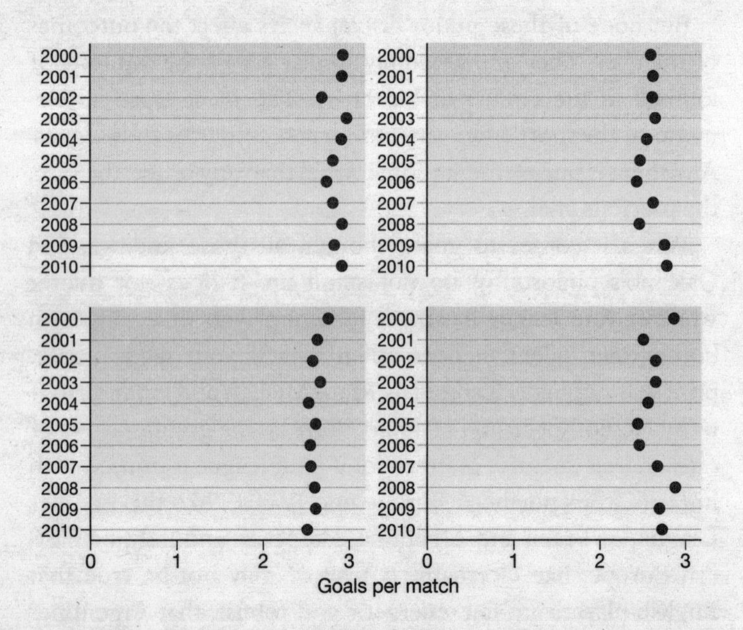

Figure 18a Goals per match in Europe's top four leagues, 2000/01–2010/11

The nature of the game is incredibly uniform at the top level. It does not matter where you are playing or where your players are from: the essential features of the game, goal production and goal prevention are as similar as you can imagine. This convergence has not occurred in lesser leagues, such as the Dutch Eredivisie, Ligue 1 in France or US Major League Soccer. Differences thrive at lower altitudes. At the game's summit, the outlook is broadly the same.

Figure 18b repeats Figure 18a, adding labels for each league. The top leagues all average slightly fewer than three goals per match, and there is little variation. The production numbers in these leagues are extremely consistent – especially considering that we are talking about a span of a decade and very different countries and leagues – and it's hard to detect any real-time trends or cross-league differences. Virtually without fail, spectators in the biggest leagues of European football saw more than two and a half and fewer than three goals per average match over the last decade – no matter where they went to the stadium.

That is not what we are taught to believe. We are informed consistently that differences in styles, tactics and personnel all matter, that in Italy the game is more defensive, in Spain it is more elegant and in the Premier League more physical, more exciting. Football's culture changes from country to country and continent to continent. We all know that.

So what about how those goals are produced? 'Tell me how you play and I'll tell you who you are,' as Galeano wrote. Surely in England the majority of goals came from inswinging crosses being met by thunderous headers, in Spain from long, flowing passing moves and in Italy from lightning-fast counter-attacks?

But here, too, features of a game that we can count – things

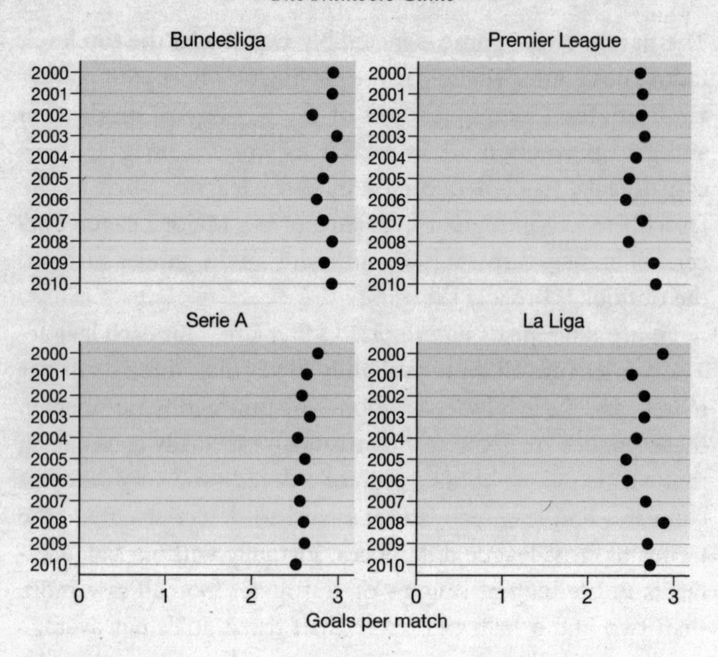

Figure 18b Goals per match in Europe's top four leagues, 2000/01–2010/11

like passes and shots – look very similar between leagues. Opta figures from the 2010/11 season show that the average team in the average match in the top four leagues in Europe completed between 425 (Bundesliga) and 449 (Serie A) passes. In Italy, only 54 of those passes were long, whereas the high was in Germany, too, with 59; those two countries provided the bookends for short passes: Germany with 332 per game, Italy 356. The differences between nations are cosmetic, shallow. The game is the same across the world's elite football leagues. If it was not for the shirts, you would not be able to tell them apart.

Table 2 Passes per game in the top four European leagues, 2010/11

	Total passes	Long passes	Short passes
Bundesliga	425	59	332
La Liga	448	56	355
Premier League	438	57	343
Serie A	449	54	356

This convergence holds true for many other key measures. The data also show that teams took roughly the same number of total shots on goal (14) and shots on target (4.7), they earned a similar number of corners (about 5), and they are awarded roughly the same number of penalty kicks (0.14) per game, too.

We also found that the number of free kicks, crosses from open play, or headed goals were pretty much the same.

Table 3 Shots, corners and penalties in the top four European leagues, 2010/11

	Shots	Shots on Target	Corners	Penalties
Bundesliga	12.9	4.6	4.9	0.14
La Liga	13.0	4.8	5.4	0.15
Premier League	14.5	4.6	5.5	0.13
Serie A	13.8	4.4	5.3	0.14

So although referees reach for their whistles and cards considerably more in Spain, and though the game may appear much faster in England than in Italy, such differences are more cosmetic than we believe. Whatever differences exist between the leagues are considerably smaller than year-on-year variations.

Stereotypes might make us think we're all different, but when it comes down to what really matters, when the game is stripped down to its basic components, we're more the same than we would like to admit. Goals are just as rare and just as beautiful, wherever the world's top players do their job.

3.

They Should Have Bought Darren Bent

Sometimes in football, you have to score goals.

Thierry Henry

The Budget Minister was furious. It was, he said, 'indecent'. The Sports Minister described it as 'deplorable', while his predecessor pronounced herself 'disgusted'. Even the President of the Republic got involved. This was not some parliamentary sex scandal, though, or an affair which had caused outright fury in France. No, this was simply the reaction to the decision by Paris Saint-Germain's Qatari owners, in summer 2012, to pay their star striker €1 million every month for four years, after tax – meaning footing a bill of €35 million every year for one player, in addition to his €25 million transfer fee.

How can a club justify spending such an exorbitant sum on one footballer, no matter how talented, even when the money is coming from an oil-rich Arab state determined to build one of the world's foremost clubs? In PSG's case, it's simple: they were not spending €165 million on a player. They were spending it on a guarantee of success.

Zlatan Ibrahimović, the player in question, is a serial title winner. Between 2003 and 2011, the giant Swedish striker won a league championship every single year, wherever he played. That's eight straight titles, including one in Holland, one in Spain and six in Serie A. He is more than just a good-luck charm: only once did he fail to score more than fourteen goals in a league season. Ibrahimović is not a passenger, along for the ride; he is a difference-maker.

It is the goals that make Ibrahimović so valuable. Indeed, it is unfair to single the Swede out when, all over the world, premium fees and astronomical wages are set aside for strikers. After all, they are the men who provide the rare, precious commodity that makes football what it is and what we all love.

Take the final day of the January transfer window in 2011: that dramatic night when Fernando Torres was flown down to Chelsea, who had paid £50 million, and presented to the club's fans just after midnight.

As his former club, Liverpool, came to terms with the loss of their idol, they, too, heard the dull thud of rotor blades above their Melwood training base. Just hours after paying a then club-record fee of £23.6 million for Luis Suárez, of Ibrahimović's former club Ajax, Liverpool lavished £35 million on Andy Carroll, the striker helicoptered in from Newcastle to complete the deal before the deadline.

Goals are rare all over the world. They are rare in games, when you consider that, on average, a Premier League team will score one goal or less in 63 per cent of its league matches, and in 30 per cent it will fail to score at all. Goals are rare for players: over three Premier League seasons between 2008 and 2011, 861 players saw playing time – a total of 30,937 individual player match appearances. The vast majority of these appearances – 28,326 or 91.6 per cent – ended with the player not scoring, 45 per cent of players never scored a single goal for three seasons and 17,322 individual match appearances – 56 per cent – ended without the player taking a single shot; a little over 80 per cent of the time, a player takes either no shots, or just one.

A quarter of all players over those three years – 221 – did not even have a shot on goal. Over three years. No shots. In three years.

No wonder, then, those select few who can not only shoot but score are so highly prized, so highly valued by football's free market; no wonder, as in the case of Ibrahimović, it is believed that there is such a direct corollary between goals and wins, and wins and trophies. Clubs pay a lot for forwards, and they pay forwards a lot, because they know quite how valuable goals are: goals win games, goals get points.

But that is not the same as saying that every goal has equal value. Some goals are worth rather more than others.

From Silver to the Gold Standard

Ibrahimović owes at least part of his salary to one of football's true innovators: Jimmy Hill. In his later years, Hill became familiar as a television presenter and pundit, but there was a time when, as chairman of the Professional Footballers' Association in the 1950s, Hill was far from part of the establishment. In many ways he was a revolutionary.

It was Hill's campaign to scrap the Football League's maximum wage – a paltry £20 a week – that led, slowly but surely, to the inflated salaries of today's Premier League stars.

Hill became chairman of Coventry in 1961, as the maximum wage was being abolished, and masterminded the Sky Blue Revolution that transformed the club; the colour of their kit changed, they sold the first-ever match-day programme and he even gave them a club song. Later, he would commission England's first all-seater stadium.

His most significant legacy, though, is the three-point rule. Hill had long thought that football had become too defensive and dull, too uninteresting for spectators. He felt instinctively that goals had become rarer with every passing season. This needed to change for professional football to thrive. Hill's solution was as simple as it was far-reaching: he proposed that victory should be rewarded with three, rather than two, points to make wins more valuable – the equivalent of shifting from silver to the gold standard. After a trial run in the Isthmian League for a few years in the 1970s, Hill won over the Football Association and convinced them to give three points for a win a trial run in 1981.

The experiment was judged such a success that, in 1995, Fifa followed suit, commanding that all its constituent leagues award three points for a victory. Sepp Blatter, General Secretary of the game's governing body, called it 'the most important sporting decision taken here, but it rewards attacking soccer'. The reward, the theory went, was 50 per cent greater, so teams would take more risks, leading to more goals, more entertainment, more fans.

To find out whether the change worked as intended should be simple: compare how many goals were scored in the season preceding the introduction of three points for a win to the number scored during the following campaign. Such an approach is insufficient, though, because such a relatively small sample size could have any number of other factors at play, from the varying qualities of relegated teams and promoted clubs to changes in ownership, coaching and even the weather. A more accurate scientific method – and an experimental mindset – is required.

Two German economists, Alexander Dilger and Hannah Geyer, came up with a way to test what changed when their nation's football leagues switched to three points for a win. They looked at 6,000 league games and 1,300 from cup competitions over the ten years before the rule change and the ten after. The cup games provided the control group, unaffected by the switch (since the reward in tournament football is progression, not points).

Dilger and Geyer did find that the three-point rule had a dramatic effect on one aspect of a football match, but it wasn't goals. In league games three points for a win led to a drastic increase in the number of yellow cards. Attacking football had increased, but the 'attack' consisted not of strikes on goal, but rather of clips of the opponents' heels, pushes in their backs, and late tackles.

There was also a clear decline in the number of draws – understandable, since losing two points for parity is less palatable than only losing one – and a rise in the number of victories by a one-goal margin.

With three points available for victory, a manager's substitutes were focused on defence, back lines refused to move forward, and the number of long clearances rose.[1] Goals had not become more abundant, but they had become even more decisive and valuable. Three points for a win had not rewarded attacking football. It had rewarded cynical football.

Here too, Ibrahimović owes Hill a little of his income. Even an artificial attempt to increase the frequency of goals failed; in a number of ways, the introduction of three points for a win may even have made goals harder to score, so many more fouls did strikers now have to endure. A player like the giant Swede who can still find the net regularly is genuinely priceless, at least to a football club. The French Budget Minister may disagree.

That is not to say that clubs should go out and lavish millions on any old forward; indeed, there is an inefficiency to how strikers are identified that could be costing teams across the world fortunes. Goals are rare and goals are valuable, but as we mentioned before, not all goals are worth the same.

The Floating Exchange Rate

We have seen that their scarcity makes goals more valuable, in terms of points, in football than in other team sports, just as goals are almost equally infrequent across football's elite. That should mean it is possible to find a uniform value for goals across the game's big leagues. And just as there is an exchange rate for turning pounds into dollars and euros into pounds, there is an exchange rate for turning goals into points.

There is one crucial difference. Unlike currency, where the rate in dollars for the first pound exchanged is identical to the rate for the eighth pound, we will see that the exchange rate for goals is entirely dependent on how many goals have already been cashed.

A straightforward way to see this is by calculating how many points an average team won per match, depending on the number of goals they scored in that match (Figure 19). And to make sure this number represents long-term tendencies, we used data from the 2000s for the four top leagues – the Bundesliga, Serie A, the Premier League and La Liga; Paris Saint-Germain, Ibrahimović's new home, do not feature just yet.

The first thing the data tell us is relatively obvious. Scoring five goals or more virtually guarantees a team all three points. There are one or two historical exceptions, outside our data set, including a pair of 6–6 draws between Leicester and Arsenal

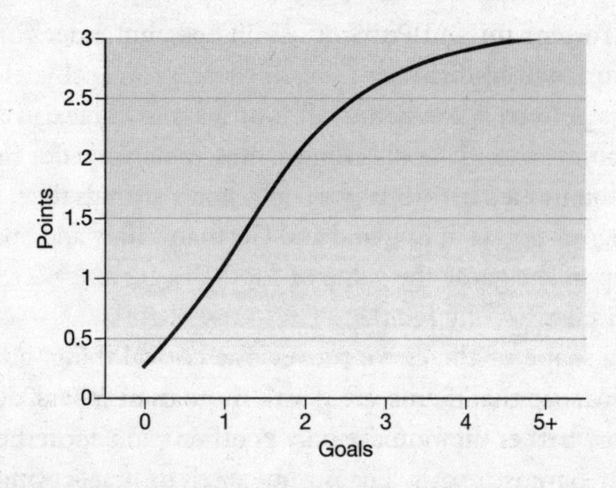

Figure 19 Goals and points per match in Europe's top four leagues, 2000–2011

and Charlton and Middlesbrough in 1930 and 1960 respectively, but the basic truth holds: as soon as you score your fifth goal, you can reasonably expect to have guaranteed victory.

It also shouldn't come as a surprise that not scoring at all doesn't yield very much in terms of points. But that is not the same as saying that scoring no goals does not glean any points at all: between 7 and 8 per cent of games end in goalless draws, so zero goals on those occasions will earn a team a point.

Those are extremes. It is in the middle of the distribution where our graph rises most sharply before levelling off. It is in this incline that goals are most valuable.

A single goal virtually guarantees at least a point, statistically speaking; two goals gets a team closer to a win than a draw; at more than two goals, teams get very close to a win, though even three or four goals do not quite guarantee victory; Newcastle, having recovered from four goals down against Arsenal and poor Reading, scorers of four goals against both Tottenham and Portsmouth in 2007 but losers on each occasion, will confirm that.

This pattern holds across all four leagues. There are slight variations – a single goal is slightly less valuable in the Bundesliga than in La Liga – but generally, goals are worth the same number of points in England and Germany, Italy and Spain. At the top of the game the value of football's currency is remarkably similar.[2]

The shape of the curve proves one critical thing: goals are not created equal. Some are worth more than others, depending on whether they are the only goals scored or whether they already have company. The numbers tell us that scoring three goals doesn't give you three times as many points as one single goal, and four goals – an increase of 33.3 per cent in goals from having scored three – doesn't give you 33.3 per cent more points than three goals do.

In other words, the exchange rate of each goal varies according to how many other goals have been scored in the game.

As Figure 20 reveals, the most valuable goal is the second (increasing the team's predicted point value by 0.99). In contrast, going from a likely thrashing to a probable stomping (that is, a fifth goal) is exchanged for only 0.1 points. This does not change between countries: two goals in Italy are worth about the same number of points as are two goals in Spain. Strikers struggling for form – as happened to both Carroll and Torres after that dramatic deadline day in 2011 – would disagree, but not all goals mean as much as each other, at least to the team's chance of success.

There are times when those extra, worthless goals later become extraordinarily important: that 6–1 rout at Old Trafford in October 2011 effectively handed Manchester City the Premier League title, on goal difference, in May 2012. These are excep-

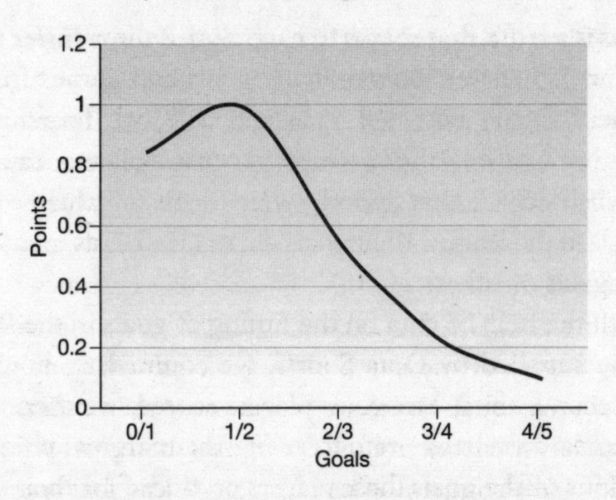

Figure 20 Marginal points produced by goals

tions, though. Teams looking to win more games need to know which players can score the goals that matter most.

Strike Price

This may seem an abstract exercise but it has some very real ramifications for the game. If a team's second goal is the most valuable, and between them the first and the second are vastly more valuable than the rest, then it suggests that the old technique of simply tallying up a striker's goals as an assessment of his productivity – and a basis for his estimated value – is simply wrong.

Strikers who score the key goals, the ones that can be directly translated into more wins and more points, are worth rather more than the flat-track bullies who appear to rub salt into wounds, scoring the third and fourth goals as victory turns into a drubbing. Simply counting strikes can be deceiving: one goal is not the same as another.

This is a truth that seems to have eluded the transfer market thus far. When we looked at the goals and games from the Premier League between 2009 and 2011, we found neither Torres nor Carroll – the two most expensive players transferred to English sides in that period – were the most valuable scorers of goals in the league; their goals did not lead to as many points as the goals of others.

With the help of data on the timing of goals in the Premier League supplied by Opta Sports, we counted up how many first, second, third, etc. goals players scored; we then applied the standard exchange rate to create the marginal points contributions of the goals these players produced for their teams.[3] By and large the rankings of the actual goals players scored and the points those goals produced lined up, but it is interesting to note that, in the lists of players who produced the most points, the Premier League's top scorer in 2009/10 (Chelsea's Didier Drogba) was third and one of the two joint top scorers in 2010/11 (Manchester United's Dimitar Berbatov) was fourth.

In 2009/10 that honour went to Wayne Rooney (though it should be noted that seven of his goals came from penalties), and in 2010/11 it went to Berbatov's former teammate and the joint winner of that year's Golden Boot, Carlos Tevez of Manchester City. So what does this tell us? Drogba and Berbatov managed to score goals when it counted for less, points-wise, to their club.

But it's not just at the top that these data are interesting. Some of the performers further down – players at teams that weren't contending for the title – were much more important to their club's fortunes than a simple tally of goals might suggest. For example, in 2010/11 Berbatov's marginal points contributions were just a notch above West Bromwich Albion's Peter Odemwingie, who scored five fewer goals than the Bulgarian. For West Brom and Odemwingie, less was actually more points-wise. The same could go for Louis Saha's tally for Everton the previous year, when his thirteen were almost as valuable in real terms as the eighteen Jermain Defoe scored for Tottenham.

The real hero of this list is Darren Bent. Indeed, if Chelsea had analysed goals using our methodology, rather than a simple count of who had scored the most, perhaps they would have realized that the way to turn around their desperate league form in January 2011 was not by splashing £50 million on Torres, but by paying half that for Bent, the most consistent marginal points producer each of the two seasons. And if Roman Abramovich had taken time to notice what proportion of his team's points were directly down to Bent's goals, his mind would have been made up. Here, too, Bent's star is in the ascendant.

When we calculated the portion of all points a club won that were due to points contributions from individual players, Darren Bent was the most valuable player both years. In 2009/10 he topped the list with 45.5 per cent of Sunderland's points, followed at some distance by West Ham's Carlton Cole at 27.9 per cent.

In 2010/11, he was again tops (if we take the points he contributed to each of the teams he played for that year – at 31.5 per cent), closely followed by Blackpool's DJ Campbell (29.7 per cent), and Odemwingie (26.7 per cent) who also managed to score at the right time and in the right order.

It is not all bad for Torres and Carroll, though. The Spaniard was ranked fifth in the league in marginal points contributions in 2009/10, but fell to eighteenth (just ahead of Wolves' Steven Fletcher and level with Sunderland's Asamoah Gyan) the following year. Carroll doesn't rank in 2009/10 (as Newcastle were playing in the Championship that season), but in 2010/11 he turns up fifteenth on the list. Perhaps when Liverpool bought him they did know what they were doing, even if subsequent evidence makes that hard to believe.

Table 4 Top 20 marginal point contributors in the Premier League, 2009/10 and 2010/11

Name	Club	Marginal Points	Goals
SEASON 2009/10			
Wayne Rooney	Manchester United	20.64	26
Darren Bent	Sunderland	20.02	24
Didier Drogba	Chelsea	19.59	29
Carlos Tevez	Manchester City	17.67	23
Fernando Torres	Liverpool	14.34	18
Frank Lampard	Chelsea	14.22	22
Jermain Defoe	Tottenham Hotspur	12.39	18
Louis Saha	Everton	11.31	13
Emmanuel Adebayor	Manchester City	10.93	15
Gabriel Agbonlahor	Aston Villa	10.86	13
Francesc Fábregas	Arsenal	10.68	15
Cameron Jerome	Birmingham City	9.77	11
Carlton Cole	West Ham United	9.75	11
Hugo Rodallega	Wigan Athletic	8.92	10
Florent Malouda	Chelsea	8.36	12
Dimitar Berbatov	Manchester United	8.34	12
Nicolas Anelka	Chelsea	8.26	11
John Carew	Aston Villa	8.04	10
Kevin Doyle	Wolverhampton	7.93	9
Dirk Kuyt	Liverpool	7.91	19

Name	Club	Marginal Points	Goals
SEASON 2010/11			
Carlos Tevez	Manchester City	15.70	20
Darren Bent	Sunderland/Aston Villa	15.01	17
Robin van Persie	Arsenal	13.60	18
Dimitar Berbatov	Manchester United	13.04	20
Peter Odemwingie	West Bromwich Albion	12.57	15
DJ Campbell	Blackpool	11.59	13
Dirk Kuyt	Liverpool	11.29	13
Rafael van der Vaart	Tottenham Hotspur	11.17	13
Javier Hernández	Manchester United	11.13	13
Clint Dempsey	Fulham	10.90	12
Charlie Adam	Blackpool	10.27	13
Florent Malouda	Chelsea	9.36	13
Samir Nasri	Arsenal	9.20	10
Wayne Rooney	Manchester United	9.17	11
Andy Carroll	Newcastle United/Liverpool	8.92	13
Didier Drogba	Chelsea	8.72	11
Kevin Nolan	Newcastle United	8.68	12
Asamoah Gyan	Sunderland	8.62	10
Fernando Torres	Liverpool/Chelsea	8.62	10
Steven Fletcher	Wolverhampton Wanderers	8.32	10

A Guide to Leaving the Stadium Early

The idea that not all goals are created equal does not just apply to the transfer market. It applies to the real business of football: winning league championships, qualifying for Europe or, at the other end of the scale, simply surviving to fight another day.

Between those two ends of the curve, though, we see again that some goals are worth more than others.

Take the first goal: we can say that scoring a single goal in every match guarantees a club it won't be relegated. Given a thirty-eight-match season in the Premier League, for example, thirty-eight goals on average will produce a points total that has been sufficient for survival in the league (43) in each of the last ten seasons, with teams having survived with as few as 34 or 35 points.[4] Survival means real money – the revenue difference between the average Premier League and Championship club in terms of television income alone is roughly £45 million, an amount that will only increase.

But if a single goal gives a team an average chance of earning at least a point, it only brings a one-in-four chance of actually being enough to secure victory. For sides with rather grander ambitions than mere survival in the league, it is the consistent second goal that is crucial.

Our figures show that it takes two goals to make the match better than a 50/50 proposition for a team, to take a side into territory where it is winning more games than it is losing (Figure 21). By the time a team manages to score three goals, supporters may be able to risk leaving early to beat the traffic with good conscience. Assuming their defence has not conceded three, by the time your team has scored four, it should be safe to slip away.

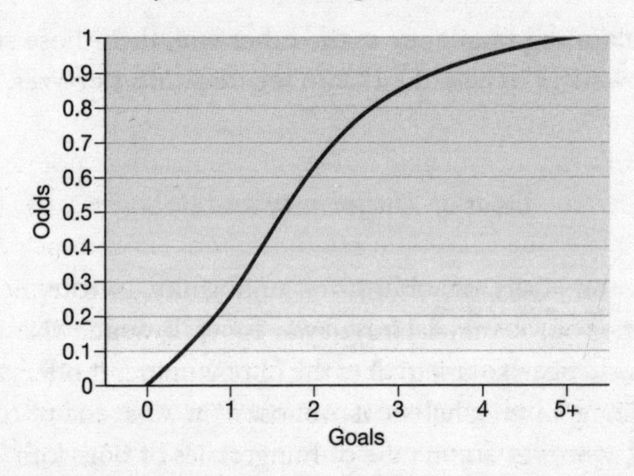

Figure 21 Number of goals and odds of match win

As with points, the connection between goals and match outcomes is not a straight line – it's S-shaped. In football, more isn't always significantly better. There may be entertainment value to a third or fourth goal, but for the objective that really matters – points and hence league position – they do not matter very much.

It is important to remember that there are two sides to every goal. They are not just scored, they are also conceded; every triumph for a striker is a disaster for a defender. These positive S-curves when looked at by the attacking side become a negative slippery slope for the defence: the first goal costs you significantly, but the second goal you let in is really the most expensive in terms of points. It may also be the one that breaks your spirit.

It is a curiosity that, with one or two exceptions such as Gianluigi Buffon and Rio Ferdinand, clubs continue to place such a premium on those who score goals, rather than those who prevent them. Strikers are fun to watch, but their contributions are only valuable because of their rarity; if goals are arriving with alarming frequency at the other end, then those strikers are powerless to help their team secure points or prizes.

Days of Uniformity and Balance

'These are days of obligatory uniformity,' wrote Eduardo Galeano, our seminal Uruguayan football writer. 'Never has the world been so unequal in the opportunities it offers and so equalizing in the habits it imposes: in this end-of-century world, whoever doesn't die of hunger dies of boredom.'[5]

Our findings would have offered him little solace. Football is essentially the same at the elite level globally. The game's culture varies tremendously from Brazil to Germany and from Ghana to Scotland, but the overall patterns of goal production look an awful lot alike in the world's most competitive professional leagues.

We suspect Galeano would roundly condemn the historical trend towards fewer goals and the similarity in top-level football. Not only have goals become an ever more precious commodity, they are produced in similar quantities by the game's very best strikers. Premier League matches see fewer interruptions and are played faster than Serie A matches, but the end result turns out to be very similar.

If he can't get diversity, Galeano wishes for beauty, regardless of its provenance: 'Years have gone by and I've finally learned to accept myself for who I am: a beggar for good football. I go about the world, hand outstretched, and in the stadiums I plead: "A pretty move, for the love of God." And when good football happens, I give thanks for the miracle and I don't give a damn which team or country performs it.'[6]

Galeano is typical of most fans. People prefer to watch a certain style of football. Some like a fast-paced, athletic style with fewer passes and lots of shots on goal, the sort of frenzied counter-attacking style employed by Manchester United or Borussia Dortmund; others prefer a systematic, deliberate build-up, circulation football, with teams maintaining possession and creating a stranglehold on the other side, like Barcelona and Spain. Both suggest fans know how much a goal is worth, and they want their team to set out to get it.

They know that goals mean survival or success. They want strikers who can produce those match-winning, season-changing strikes with regularity, they want their chairmen to spend a fortune to bring them to their club, and they want their managers to set their team up to give them as many chances to score them as possible.

The history of football is the history of the goal. How it grew ever more rare and ever more precious until it reached what appears to be its base rate in recent years, and how those who could provide it became increasingly valuable, ever more revered, how teams endeavoured to find ways of scoring more and conceding fewer. It is that search – for more at one end and fewer at the other – that has prompted one hundred years of tactical insight and innovation and that has made football what it is today: not the attacking game, but a balance between two opposing forces. A sport of light and dark.

On the Pitch: Football 'Intelligence' and Why Less Can Be More

4.
Light and Dark

We play leftist football. Everyone does everything.

Pep Guardiola

Football's rich and illustrious history is full of philosophers, preachers and proselytizers, but few have ever looked the part of visionary quite so much as César Luis Menotti, the wild-haired, chain-smoking manager of Argentina's 1978 World Cup winners.

Menotti, known as *El Flaco* – 'The Thin One' – had the larger-than-life, intellectual personality to match his idiosyncratic look. A lifelong Communist who took charge of his national football side when the country was ruled by a brutal, right-wing military junta, Menotti was, when it came to his career, something of a pragmatist. He would, though, resent such a charge if it were levelled at his beliefs about football. On the pitch, Menotti was a purist.

His message was simple: football is about scoring one, two, or three more goals than the opposition. He was not interested in securing a lead and then shutting up shop. We have seen that the game, the modern game, is about balance. But to Menotti, there were no shades of grey. There was attacking, dazzling and exciting, and there was defending, cynical and miserable. There was light and there was dark.[1]

Menotti treated this as a difference in ideology. He spoke of 'left-wing' football and 'right-wing' football; to Menotti the Communist, Menotti the purist, the former was positive, marked by creativity and joy, while the latter was negative, fearful, defined by an obsession with results. 'Right-wing football wants to suggest that life is struggle,' he said. 'It demands sacrifices. We have to become of steel and win by any method . . . obey and function, that's what those with power want from the players. That's how they create retards, useful idiots that go with the system.'[2]

In truth his sides were always a little more systematized than he would like to admit.[3] He too, when all was said and done, had his contradictions (as a Marxist who did business with a murderous right-wing junta). That does not alter the fact that his ideas are seductive. He counts Jorge Valdano, the long-time Technical Director of Real Madrid, and Jürgen Klinsmann, the former Germany manager and current coach of the United States, among his acolytes; his principles are no doubt shared by the likes of Cruyff, Pep Guardiola, Arsène Wenger, Marcelo Bielsa, Zdeněk Zeman, Brendan Rodgers and even Ian Holloway.

Most fans find themselves broadly agreeing with his idea that attacking is to be encouraged, and defending is a last resort. That is why attackers are so prized – by the transfer market, by clubs, by those who hand out individual awards at the end of the season, by the people who edit highlights reels – while defenders are undervalued, financially and otherwise. If the ultimate aim of football is the goal, then we should do all we can to set out and get it.

But is the Argentine's idea that a strong attack can pick a way past even the most tightly massed defence realistic? How can we know whether the contention is true? Well, Menotti's approach can be viewed as a theory; his idea that scoring more is better than conceding less is just a hypothesis. And, like any hypothesis, it can be tested against data. When put to the test, will Menotti's ideas hold water or will they, to quote the British biologist Thomas Huxley, fall victim to 'the great tragedy of science, the slaying of a beautiful hypothesis by an ugly fact'?[4]

In football there is no more beautiful hypothesis than the idea that attack will always win out. Millions of pounds ride on it every season, as teams clamber over each other to sign world-class forwards, paying them ever more eye-watering salaries. These are the stars of the sport, after all; the men who can make the difference between success and failure. Have the best attack, and no defence will be able to stop you on your quest for glory. So goes the rough logic. Football, as we have seen, is the goal, and the goal is football.

But should we be more concerned with scoring them or averting them? Should we demand the club we support spend more on buying an additional forward or another centre back? For more than a century, those who have thought about and played the game have favoured the former. Is this approach actually correct? Is one really more valuable than the other? Are we playing the game the right way?

To Win or Not to Lose?

Let's interrogate the data properly, the way a good attack would ask questions of a well-organized defence. We collected twenty years' worth of results across the top four European leagues. The first question we asked was this: do teams that score the most goals always win the league?

The simple answer is no. On average, teams that scored the most goals in a season won only about half (51 per cent) of the championships available – ranging from a low of eight of all twenty Bundesliga seasons to a high of twelve in the Premier League. Scoring the most goals does not guarantee a championship – far from it.

And what about the darker side of the force? Do the teams that concede the fewest goals win titles? Again, not necessarily. The best defence will pick up a championship 46 per cent of the time, with the range from a low of 40 per cent in the Premier League and La Liga to a high of 55 per cent in Italy (Figure 22). Scoring the most goals over the course of a season gives you slightly better odds of winning the league than conceding the fewest, but as a strategy for all but guaranteeing a championship it seems to fall well short.

These are not two ways of saying the same thing, by the way: the clubs scoring the most goals often weren't the ones conceding the fewest. Of the eighty champions included in our data (twenty seasons across four leagues), only sixteen were their league's best at both ends of the pitch.

Winning a title can come down to the finest of margins – witness Manchester City's last-gasp Premier League victory in 2012 – so this is far from conclusive. A better approach may be to see whether league position is more strongly associated with goals scored or goals conceded. If there is a tighter statistical connection between where you finish and how many you score, then Menotti and his followers would seem to be right; if there's a better fit with how few you concede, perhaps the Argentine's 'right-wing' football isn't as stifling and miserablist as he believes.

Figure 23 on the following page shows Premier League data from 2001/02 to 2010/11 of goals (scored and conceded)

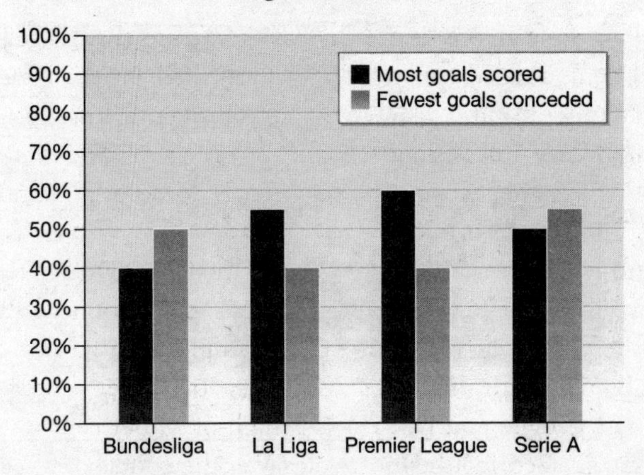

Figure 22 Offence, defence and percentage of seasons won, 1991–2010

and points won by the clubs that competed across the seasons.[5]

There are two strategies for producing points in the Premier League: you get more points if you score more goals, but conceding fewer is equally effective. The steepness of our trend lines is similar and both sets of points cluster tightly to those lines. These numbers do not prove Menotti right, nor do they suggest he is wrong. But they do suggest there may be more to the story than siding with one or the other. Maybe football is a sport of shades of grey.

There's one flaw with this technique: it does not show us *how* teams won these points. They could have done it by winning games, or they could have done it by avoiding defeat; thanks to Jimmy Hill, one win and two losses produce the same number of points as three successive draws.

Which of those two options you prefer may well say a lot about your approach to football – whether you are one of

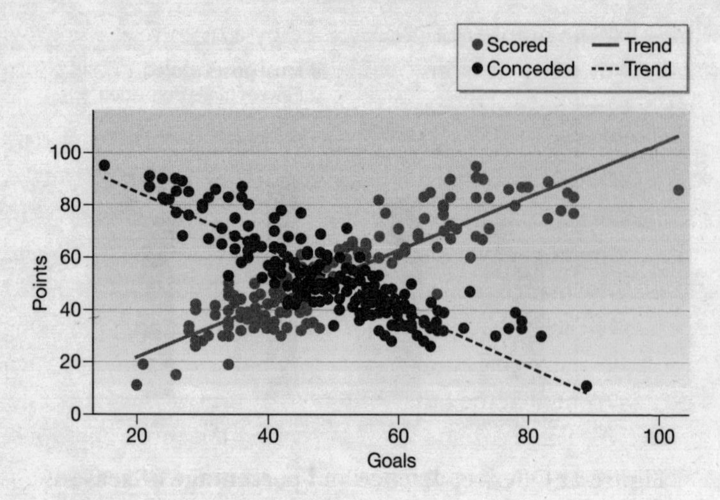

Figure 23 The relationship between goals and points, Premier League, 2001/02–2010/11

Menotti's scorned right-wingers, or whether you join him on the left. Would you rather your team tasted victory once, and then had to endure defeat twice, or is it better that it does not lose at all? We know which way Menotti would go but there are others – José Mourinho, for one – who would sacrifice glory to avoid the ignominy of defeat. What we want to know is not which one is prettier, or morally superior, but something rather more rudimentary: is it better to win, or not to lose?

To see if attacking leads to more wins, and whether defence leads to fewer wins and more draws, we conducted a set of rigorous, sophisticated regression analyses on our Premier League data. This technique allows us to see if we can predict a team's results based on information about its combined defensive and attacking performance and to judge whether one is a more powerful tool than the other. The key is that we can then examine how the number of goals scored by a team relates to how many matches it wins, while simultaneously taking into account its defensive record (and vice versa).

The regressions give us factors (coefficients) that translate an additional goal scored and a goal not conceded into a share of a win or a loss. These are more complex versions of the exchange rate we saw earlier: regardless of a team's defensive performance (conceding X number of goals), how many points is each additional goal scored worth? And controlling for how many goals a team scores, what is the value of each goal prevented at the other end?

Between 2001/02 and 2010/11, scoring ten more goals over the course of a season was worth, all else being equal, an additional 2.30 wins, while conceding ten fewer goals was worth 2.16 additional wins in the Premier League. That means goals created and goals prevented contribute about equally to manufacturing *wins* in English football.

It is when we look at the number of games a club could expect to *lose*, though, that goals scored and goals conceded begin to vary in significance. A good attack, like a good defence, decreases the number of losses a club racks up, but defence provides a more powerful statistical explanation for why teams lose.

How much more powerful? *Scoring* an additional ten goals reduced a club's expected number of defeats per season by 1.76; *conceding* ten fewer goals reduced defeats in the Premier League by 2.35 matches. So when it came to avoiding defeat, the goals that clubs didn't concede were each 33 per cent more valuable than the goals they scored.

What does all this tell us? It shows that Menotti was *wrong* in thinking that attack was a recipe for success by itself; attack and defence matter equally for climbing the final league table by May. You're more likely to win a title or avoid relegation if you have a better back line, regardless of how many goals your strikers can produce.

Simply trying to win games on the back of a good attack is not enough to take a team to glory. You have to not lose them. Neither a left-wing team nor a right-wing team has a perfect recipe for success; the goal is somewhere in the middle.

We See a Game

Daniel Alves may be one of the finest right backs on the planet, but it should be no surprise that a Brazilian playing for Barcelona falls on the left-hand side of Menotti's politics of football.

'Chelsea,' the shaven-headed full back said of the team Pep Guardiola's men had beaten to a place in the Champions League final of 2009, 'did not reach the final because of fear. The team that has got a man more, is playing at home and is winning should have attacked us more. If you don't have that concept of football that Barcelona have, you stay back, and you get knocked out. You have to go forward. Stay back: losers. Go forward: winners. Chelsea lacked the courage to take a step forward and attack us. At that moment, we realized they had renounced the game.'[6]

Stay back: losers. Go forward: winners. Alves is not alone in his stark assessment of football. There is a right way and a wrong way to play the game, and the right way will always bear out. This contrast dates back to the very earliest days of organized football: a piece in the *Scottish Athletic Journal* of November 1882 roundly condemned the habit of 'certain country clubs' of keeping two men back twenty yards from their own goal. Defending, even then, simply was not the right way to play the game; the sport was supposed to be about all-out attack, attempting to outscore the opposition.

This early imprint on football left a powerful legacy that has continued to affect how we see the game. Italy's perfection of *catenaccio* is used as a stick with which to beat Serie A as dull, defensive; Greece's triumph in Euro 2004 was not exactly celebrated outside Athens. (And, we suspect that even results-minded Italians and Greeks would prefer to win by attacking than by not conceding.) Where attacking play is lionized, impressive defences are shunned. Strikers attract the fat fees and the high salaries, and win awards and hearts; centre backs are condemned to toil in relative anonymity, if not relative penury.

That is true in Argentina, as it is around the globe; the country's footballing motto is best expressed as *Ganar, gustar, golear*: to win, to delight, to thrash. *La Nuestra*, the Argentine vision of football, concentrates on the art of dribbling and a dash of trickery; it is held to be more individual than the game played in Europe. No wonder Menotti was so enamoured of attack. His footballing culture, just like all our footballing cultures, compelled him to be.

There is nothing inherently wrong with that. Most of our favourite memories of football are of flowing moves and wonderful goals; most of us idolize George Best or Lionel Messi rather than Bobby Moore or Carles Puyol. But football's obsession with attack does have one negative consequence: the role played by defence, and defenders, is underestimated and misunderstood. Remember our earlier discussion of the dismal performance of defenders and goalkeepers in the Ballon d'Or balloting. There are deep psychological reasons for this; reasons that give us an explanation for why we remember the goals that were scored more than those that were not, and, by extension, why we believe that attack is more important, more worthy, than defence, even though the numbers suggest that is not the case at all.

At the most basic level, there is the Hedonist Principle, which assumes people will seek pleasure and avoid pain to satisfy their basic biological and psychological needs. Football is a game that has long associated scoring with winning and vice versa, and so putting the ball in the net means immediate pleasure; preventing someone else from doing so denies them that same joy. All of football's positive emotions go with attacking: creating, conquering, overcoming, releasing. Defence is inherently negative, repressive, playing to avoid defeat.

We remember the positives much more easily. This is to do with what psychologists call 'decision bias' and 'motivated reasoning'. We are hard-wired to reach biased interpretations of data that run counter to beliefs we hold and care deeply about. So when we are called on to examine objective evidence or information, we are predisposed to look at the evidence that supports what we already believe. We see what we expect to see, and we see what we wish to see. This makes collecting and interpreting football information particularly difficult, given the tribal loyalties we have.

In a 1954 study aptly titled 'They Saw a Game', Albert Hastorf and Hadley Cantril investigated how people 'saw' what happened in a game of (American) football between Dartmouth College and Princeton University.

The game had been played in 1951; Princeton won what had turned out to be a rough contest with lots of penalties for both sides. The game had been controversial because the Princeton quarterback, an academic standout playing in his last college game, had to leave the field in the second quarter with a broken nose and a concussion. In the third quarter, the Dartmouth quarterback had to leave the field with a broken leg after another brutal tackle.

Hastorf (on the faculty of Dartmouth) and Cantril (a Princeton professor) asked spectators what, exactly, had happened. The game had been filmed, and the professors made their subjects watch it once again before questioning them about what they thought had taken place, and who they considered was to blame for the game turning ugly.

Not surprisingly, the answers varied. Even immediately after watching the game only 36 per cent of the Dartmouth students but 86 per cent of the Princeton students said it was Dartmouth who had started the rough play. In contrast, 53 per cent of the Dartmouth students and 11 per cent of the Princeton students said that both teams were at fault. When asked if they thought the game had been played fairly, 93 per cent of Princeton students thought it was rough and dirty but fewer than half (42 per cent) of Dartmouth students agreed with them. Princeton students also thought they saw the Dartmouth team make over twice as many rule violations as were reported by Dartmouth students.

Clearly, the 'facts' that people 'saw' depended on whether the observers were motivated to view one or the other side in a more positive light. As Dan Kahan, a professor at Yale University's Law School, explains about Hastorf and Cantril's classic study, 'the emotional stake the students had in affirming their loyalty to their respective institutions shaped what they saw on the tape . . . The students wanted to experience solidarity with their institutions, but they didn't treat that as a conscious reason for seeing what they saw. They had no idea . . . that their perceptions were being bent in this way.'[7]

This happens all the time, of course: English fans of a certain vintage swear the third goal in the 1966 World Cup final crossed the line, but Germans are less convinced. To some, Cristiano Ronaldo is an artist who gets fouled a lot; to others, he is a diving con man. Our brains see what they wish to see, and once we believe what we believe, we are not for moving.[8]

Tom Gilovich, a psychologist at Cornell University, knows exactly how this works. He studies how people process information and make decisions. He was the co-author of one of the most famous sports studies ever published, 'The Hot Hand in Basketball: On the Misperception of Random Sequences'. The paper revealed that there is no such thing as 'the hot hand', basketball terminology to describe a player who is in a rich vein of form. 'Streak shooting' in basketball, therefore, is a powerful myth:[9]

> Basketball players and fans alike tend to believe that a player's chance of hitting a shot are greater following a hit than following a miss on the previous shot. However, detailed analyses of the shooting records of the Philadelphia 76ers provided no evidence for a positive correlation between the outcomes of successive shots. The same conclusions emerged from free-throw records of the Boston Celtics, and from a controlled shooting experiment with the men and women of Cornell's varsity teams. The outcomes of previous shots influenced Cornell players' predictions but not their performance.

In basketball, as in many sports, a player having consecutive successes is said to be on fire, and everyone involved – the player himself, his opponent, his teammates, fans and referees –

can feel in their bones that he is on a hot streak. Gilovich et al.'s numbers proved that this feeling is simply and absolutely dead wrong. In fact the streaks that shooters have during games or in practice are identical to the sequences that arise based simply on the player's average rate of making baskets. So, for a player who hits 50 per cent of his shots, his pattern of makes and misses will be identical to the runs of heads and tails that arise when flipping a coin.

Even though the research is straightforward and the findings have been replicated a number of times, the paper created a furore in basketball circles – everyone who's anyone just 'knows' that guys 'get into a rhythm' – and the paper's findings continue to be debated by sports fans and analysts the world over. People just did not want to believe the study's results.

Gilovich is sanguine about the reception his work received, even from basketball greats like Red Auerbach. Auerbach, voted the greatest coach in NBA history and an icon for the team Gilovich supports, the Boston Celtics, was unimpressed with the study. 'So he made a study,' he replied laconically. 'I couldn't care less.'[10]

Gilovich admits that such a reaction is typical. 'Since I'm a Celtics fan, of course I wanted Red to like my work more than that,' he told us. 'But over time I've come to be very fond of the dismissal because it reinforces the message of the research – that the belief in the hot hand is a cognitive illusion and so those most closely associated with the game will have "seen" the most evidence of the hot hand and therefore be most resistant to our findings.'

Those involved in sports see what they want to see, what they are taught to see, and what they believe they see. They see a game. Auerbach just 'knows' that streak shooting exists, even though it doesn't, and we all see attack overcoming defence in football, even though it doesn't.

The Maldini Principle: Dogs That Don't Bark

Even Sir Alex Ferguson, the most successful manager in British history, is occasionally susceptible to cognitive illusions. In August 2001 the Scot decided to sell the Dutch international defender Jaap Stam to Lazio. 'The move surprised everyone,' wrote Simon Kuper. 'Some thought Ferguson was punishing the Dutchman for a silly autobiography he had just published. In truth, although Ferguson didn't say this publicly, the sale was prompted partly by match data. Studying the numbers, Ferguson had spotted that Stam was tackling less often than before. He presumed the defender, then twenty-nine, was declining. So he sold him.'[11]

Ferguson has called the decision the biggest mistake of his career. No doubt to some the story would serve as a warning as to the dangers of reducing football to a stream of numbers; to us, though, it simply proves that defence is not just undervalued in football, but valued entirely incorrectly. This is because of another psychological phenomenon that gets in the way of understanding defence: we remember, and place undue significance on, things that do happen while ignoring those that do not. As the psychologist Eliot Hearst explains: 'In many situations animals and human beings have surprising difficulty noticing and using information provided by the absence or non-occurrence of something . . . Non-occurrences of events appear generally less salient, memorable or informative than occurrences.'[12]

As a result people discount causes that are absent (things

that didn't happen) and augment the importance of causes that are present (things that did happen).[13] This influences how we think about football: not only do we consider the goals that our team score more important than the goals they do not concede, but we value the tackles they make more highly than those challenges that their preternatural sense of positioning, their game intelligence, mean they do not need to make. That is where Ferguson went wrong. He needed to engage in counterfactual thinking: Stam was not doing as much, but that was not a sign of weakness, it was a sign of his quality. But because Ferguson could not see those unmade tackles, he did not value them.

Xabi Alonso, the Spain and ex-Liverpool midfield player, understands this instinctively. He told the *Guardian* that he was surprised to see so many young players at Liverpool herald 'tackling' as one of their strengths. 'I can't get into my head that football development would educate tackling as a quality, something to learn, to teach, a characteristic of your play,' he said. 'How can that be a way of seeing the game? I just don't understand football in those terms. Tackling is a [last] resort and you will need it, but it isn't a quality to aspire to, a definition.'[14] To Alonso, tackling happens when something goes wrong, not right.

There was no greater exponent of this than Paolo Maldini, the legendary former captain of AC Milan and Italy. Maldini, famously, rarely made a tackle. Mike Forde, Chelsea's Director of Football Operations, reckons Maldini made 'one every two games'. Maldini never had to get his legs dirty because he was always in the right place to cut off the danger. The best defenders are those who never tackle. The art of good defending is about dogs that do not bark.

This is difficult to accept – even for Ferguson – because it requires us to engage in counterfactual thinking – that is, we need to imagine a world that is counter to the facts, a world that does not exist.

Tom Gilovich, the myth-busting basketball psychologist, suggests that counterfactual thinking is hard because of the way people form causal explanations for events. As a general rule, when trying to explain an outcome we see in the world, people tend to think harder about things that happen than things that don't.

Gilovich had a trick up his sleeve to make the point. Look at Figure 24. In the upper part, try to find the oval *without* a line through it (the Q that's really an O); in the lower, try to find the oval *with* a line through it. In the first, locating the O is tricky; in the second, finding the Q could not be simpler.[15] It is easier for us to find something that does exist – the dash – than something that does not. This means that, when we reason about the effects of, say, tackles made versus those that didn't happen, the absence of something is very different from the presence, and it trips us up.

This same phenomenon comes into play during penalty shootouts: scientists have found that the more anxious a player is, the more likely he is to look at the goalkeeper – something that is there – rather than the space around him.[16] Players who are told not to shoot within the keeper's reach are even more likely to look at him, an effect known as an ironic process of mental control, when the effort *not* to do something makes doing it even more likely.[17]

This bias towards seeing what is there and ignoring what is not makes valuing defence difficult. Attacking has one simple best outcome: a goal. But defending is quite the opposite: there, the best outcome is a goal that is not conceded, an event that does not actually happen. That may be because of a shot

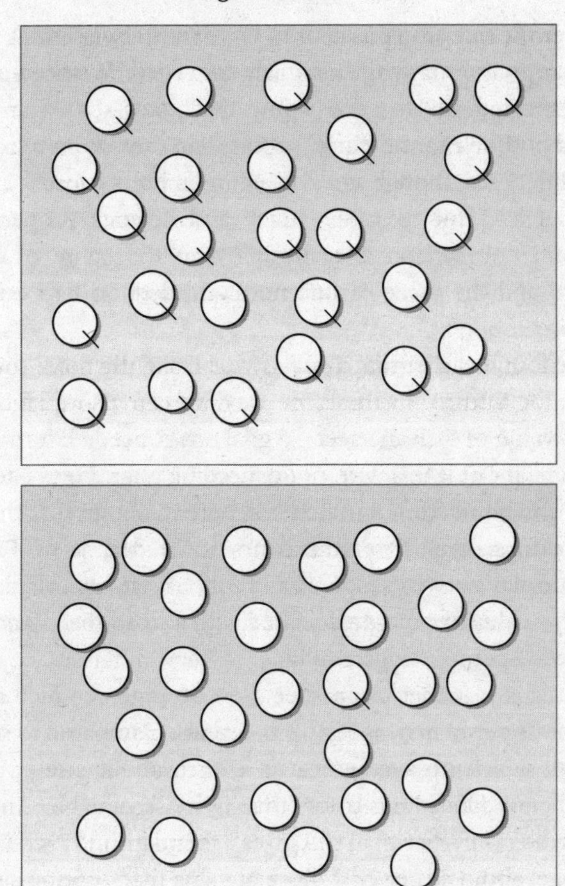

Figure 24 Absence vs presence

that did not come or a cross that was not made or a through ball that could not be weighted properly. No wonder defenders don't win the Ballon d'Or.

There is more to it, and it matters for football analysis. To answer the question posed of us by Menotti, we cannot simply look at goals scored against goals conceded. We need a more sophisticated analysis. We know that goals scored and conceded both matter to teams' success and they do so to roughly equal degrees, though not conceding matters more for avoiding defeats. But to value attack and defence properly, the relevant comparison is really between the value of a goal scored and the value of one not conceded. So let's compare the two.

We found earlier that a goal is worth slightly more than one point for a team. In the same way we can also quantify the point value of a clean sheet – a goal not conceded. It may help to think about it this way: not conceding guarantees a team at least one point from a match and potentially gives it three (if the team scores). Over the course of a decade of Premier League play between 2001/02 and 2010/11, we can calculate the average value of points associated with a clean sheet (and goals conceded per match generally).

It turns out that clean sheets on average produce almost 2.5 points per match, as Figure 25 reveals. Compared to scoring a goal, which on average earns a team about one point per match, not conceding is more than twice as valuable. And even conceding only one goal still gives a team around 1.5 points on average, about 30 per cent more in value than scoring a one.

Another way to think about this is to ask how many goals a team needs to score to generate the points produced by a clean sheet. The answer for the Premier League is 'more than two' – as the graph shows, a clean sheet produces almost as many points for a team as scoring two goals does. The numbers for the other top leagues aren't very different. In top-level football, a clean sheet or zero goals conceded is more valuable than scoring a single goal. To put this in *Numbers Game* terms,

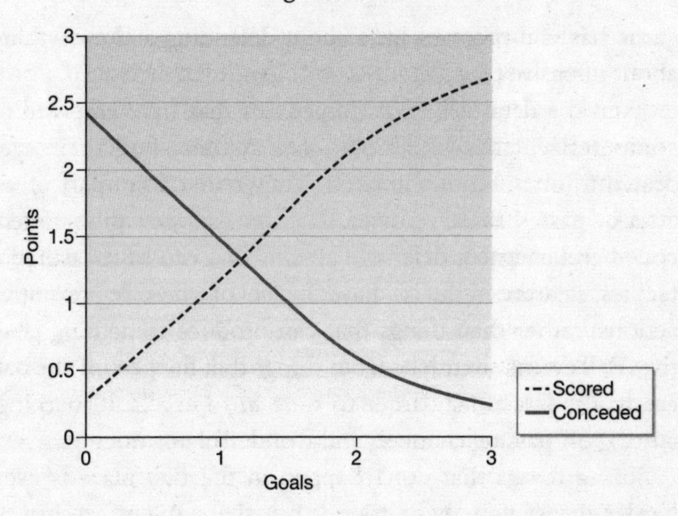

Figure 25 Point values of goals scored and conceded, Premier League, 2001/02–2010/11

then, an inequality central to understanding football is this: 0 > 1. Goals that don't happen are more valuable than those that do.

Yin and Yang

Defence has, for too long, been ignored by those who analyse and assess football. It has not been invited to play in the numbers game. Charles Reep may not have preached the same sort of style of play as Menotti, but both were blind to the fact that football is a game of light and dark, of attack and defence. Reep focused only on what it takes for teams to score, just as Menotti preached that attacking will triumph over the bleak pragmatism of defence. The debate over how best to play the game has said precious little about defending and everything about offensive play.

Even the data collection companies that have emerged to computerize traditional notational systems find their eyes drawn to one end of the pitch. Things that form part of an attack – passes, assists, crosses, shots, goals – are easily spotted, coded and counted; defensive actions that can be measured – tackles, clearances, duels – have the feel of one-offs, preventive actions, rather than things that can produce something positive. Ball events are tracked, but things that happen off the ball are ignored. It is far harder to tune in to excellent marking, cutting off passing channels and wonderful positioning.

Seeing things that don't happen in the first place is even harder to get your head round. But these things are just as important as those things that can be seen, and can be measured, if not more so.

Clean sheets are valued and mentioned, but goals are celebrated, despite the fact that 0 > 1. Strikers are loved, defenders respected. And the penchant for attack continues to be deeply embedded in how decisions are made at the top level of football. There may well be a different explanation, but we strongly suspect that goalkeepers and defenders are less likely to become managers of the world's top clubs simply because defence is neither well understood nor highly valued. Of course, there is self-selection, too: exhibitionists are drawn more to the death or glory of the forward line than the dourness of the back four, and thence thereafter to the celebrity of a life in managership over less glamorous but more secure options. In the 2011/12 season, for example, not a single Premier League manager had been a goalkeeper in his playing days, and only five of the twenty clubs had a former defender as their manager at the end of the season.

So football is a schizophrenic game: it's as much about not losing as it is about winning but pretends otherwise. Why is that? Some football cultures value artistry over results. But by definition this means that winning and losing are secondary. Historically, Germans and Englishmen saw this as a foolish approach, something adopted by unpredictable, indolent Latins.

Football is not alone in its neglect of understanding and valuing defence. As Bill James, the godfather of baseball statistics, pointed out: 'Defense is inherently harder to measure. And this is true in any sport. In any sport, the defensive statistics are more primitive than the offensive statistics. It's not just sports. It's true in life. It would be true in warfare and true in love.'[18]

This means that we are allowing our selective memories and perceptions to get in the way of a truly rational understanding of football. Menotti's professed choice between left-wing and right-wing football is a false one. Teams that score more than their opponents will always win, but so will teams that concede fewer. As Johan Cruyff said of the Italians: they can't beat you, but you can lose to them.

Your attack, in other words, is only as valuable as your defence will allow, while your defence is only as valuable as your attack makes it. Going for the jugular might be more popular, more entertaining, but there must be harmony between football's two sides. There must be yin and there must be yang, the ancient Chinese symbol of balance, interplay, contradiction and coexistence. There is defence within attack – Barcelona's *passenaccio* – and attack within defence – pulling the opponent forward to create the counter-attack, using their own ambition to weaken the other side. As Herbert Chapman long ago observed: 'A team can attack for too long.'

Menotti's advocacy of left-wing purity is both truthful and

slightly deceitful. His sides were not quite so free-flowing as he wanted you to believe, and he confesses: 'I play to win as much or more than any egoist who thinks he's going to win by other means.'[19] He is a man of contradictions: an advocate for imbalance in both life and football, and yet a practitioner of left and right, attack and defence.

That is not a criticism. The truest path, in football, lies in the middle way. If Menotti was as ideologically pure as he had wanted us to think he was, he probably would not have won the World Cup; he probably would not be quoted as one of the sport's great thinkers.

Our memories and our minds may let our eyes trick us into placing greater significance on what we can see, but it is dangerous to overvalue attack at the expense of defence. Yes, one goal for is greater than not scoring, 1 > 0, but keeping a clean sheet is more valuable than scoring a single goal, 0 > 1. Those multimillion-pound strikers are only worth investing in if your back line is solid.

Guardiola's redefinition of leftist football that begins this chapter is correct. Everyone must do everything. We must not be blinded by the light; for a team to be successful, we must pay heed to the dark, too.

5.
Piggy in the Middle

Without the ball, you can't win.

Johan Cruyff

If we have the ball, they can't score.

Johan Cruyff

Sepp Herberger was never short of a maxim. The legendary coach of the West Germany team that overcame Hungary's Magic Magyars to produce the Miracle of Bern and win the 1954 World Cup had a fine line in simple, instructive aphorisms. Many survive to this day; some have passed into cliché. Herberger is the man who coined the phrase 'the next opponent is always the hardest'.

His most famous dictum, though, was to do with the ball. The ball formed a core part of Herberger's thinking.[1] He knew that understanding the ball is central to understanding the game. The ball, as he saw it, 'is always in better shape than anyone'; the 'fastest player', he believed, 'is the ball'. His most famous quote is even simpler. It is so obvious that if anyone else had said it they may have been mocked. Having a World Cup on your résumé tends to help avoid such a fate. 'The ball,' Herberger used to say, 'is round.'

To Herberger, that phrase was a useful way of reminding fans, players, journalists and his employers that football is a game of the unexpected. Or rather the original quote was. His axiom has been abbreviated over the years, but it's worth knowing in full. His words were not just 'The ball is round,' but 'The ball is round, so that the game can change direction.' When the ball is in play, he meant, anything can happen.

Football is the goal. The game is defined by its end product. Each side possesses a light side, seeking the goal, and a dark side, hoping to divert it. And at the centre of that collision between the positive and the negative, the yin and the yang, is the ball. One side has it, the light, and one side, the side that does not, remains in the dark. To understand the game, as Herberger knew, we must understand the ball: what it means to have it, and what it means to be without it.

In recent years it has become fashionable to want to retain the ball. There are teams who almost seem to keep possession of it for its own sake, teams who want to bask in its light as much as possible. Barcelona and Spain are the most notable exponents. They treasure the ball, cherish it, and it has duly rewarded them, with Spanish league titles, with the Champions League trophy and with the championships of Europe and the world.

Plenty of other sides are just as enamoured of the ball, though, and in very different ways. It is beloved of Arsenal, of course, and the club's manager Arsène Wenger, who drastically changed the team's style after taking over from the more cautious, direct George Graham in 1996. 'Arsène Wenger's training is all about possession of the football, movement of the football and support of one another,'[2] explains Nigel Winterburn, who played under both managers.

Such a system was beloved of Brendan Rodgers's Swansea.

But ask Arsenal's French manager whether he sees similarities in the two styles of play and he will dismiss it out of hand: Swansea, to Wenger, engage in what he terms 'sterile domination', the endless recycling of possession, sweeping mandalas painted on the pitch to no end or purpose. Bayern Munich, under Louis van Gaal, were accused of the same thing. Possession for possession's sake, circulation football, an addiction to the light.

And then there are those teams who do not seem to want the ball, who are happy to spend most of their lives in the dark. There are the counter-attacking units of José Mourinho and Portugal, or the frenetic, swarming teams of Zdeněk Zeman and Antonio Conte and Jürgen Klopp's Borussia Dortmund. It is possible, as in the latter cases, to be attractive without dominating possession. There is true beauty in the dark. And there is ugliness, too, the charge often levelled at teams like the Wimbledon of the 1980s, Graham Taylor's Watford or, more recently, Tony Pulis's Stoke. These are the *wilful* have nots: the sides who have made a virtue, an art form, out of not having the ball.

The contrast between the two styles is stark. Let's take Arsenal and Stoke, teams at opposite ends of the modern Premier League possession spectrum. According to Opta Sports, over the course of the 2010/11 season, for example, Arsenal players had almost 30,000 touches of the ball.[3] They topped the league with 60 per cent possession in the average match, never had less than 46 per cent, and frequently achieved more than two-thirds of possession in a match.

Stoke, on the other hand, in the same season, saw their players touch the ball 18,451 times – the lowest in the league – and have an average of 39 per cent possession. When the two sides met at Stoke's Britannia Stadium that year, in fact, the home team had just 26 per cent of possession.[4] Stoke were only marginally more possessive of the ball on other occasions; only once that entire year did Stoke have more possession of the ball than their opponents.

There are plenty of managers out there who make light of such statistics, and we suspect Pulis is among them. Having more possession of the ball is no guarantee of victory. In fact, that day in May when Arsenal visited the Britannia and enjoyed almost 75 per cent possession – completing 611 passes to Stoke's 223 – they lost 3–1.

That is far from an isolated example. Take Barcelona, widely regarded as the finest club side in the world, contriving to lose on aggregate to Chelsea over two legs in the 2012 Champions League semi-finals. Pep Guardiola's side, brimming with the talents of Lionel Messi, Xavi Hernández, Andrés Iniesta and the rest, had 79 per cent of the possession in the first leg and 82 per cent in the second. They won neither match. It was the same that season against Mourinho's Real Madrid: Barcelona had 72 per cent of the ball, and lost. The ball is round, as Herberger would say. The unexpected does happen.

It would be comforting to chalk those results up to chance or the law of large numbers. We have seen already what a powerful factor fortune can be when it comes to football and that anything can happen if you play football often enough. We also know that, roughly half the time, the better side does not win. But we cannot just accept that sometimes the best teams lose simply because of the vicissitudes of fate. We need to establish whether, in these cases, they lost despite having all that possession or – as Herbert Chapman might suggest – because of it. Is it possible that the artists are wrong and the artisans right: can possession be worthless unless you do something with it? Is keeping the ball a means to an end or an end in itself?

To find out, there is one thing we have to do: we have to establish what being 'in possession' means. It is one of those football phrases that trips easily off the tongue; one of the rare football numbers that is discussed on television and radio, in pubs and bars, considered vastly important in determining how well a team has played or describing its characteristics. In the age of Barcelona and Spain, possession is all the rage. But what does being in possession actually mean? Once we've answered that, we can start to work out just how valuable possession is.

Chasing the Ball

First things first: let's define possession. A dictionary would have it that possession is the state of 'having something'. That is, to possess something means to have practical or physical control over an object. In the football sense, that means having control over the ball, that inflated sphere with a circumference of 68–70 cm (27–28 in) and a weight of 410–450g (14–16 oz), and doing so with your feet.

That sounds simple enough. Throw in the biomechanics involved, though, and the idea that anyone ever truly has possession over the ball becomes a little less straightforward. The ball, as Herberger noted, is round, and that's a bit of a problem: human feet are not really designed to have control over anything, let alone something spherical, reasonably big, and relatively heavy.

We can see how difficult it is for clubs in the world's most popular football league to 'possess' the ball by looking at the routes the ball travels during an ordinary match, with the help of Opta data. We took a random ten-minute span from a random Premier League match to show you – the game between Aston Villa and Wolves on 19 March 2011 (Figure 26). The ball cannons around the pitch, the haphazard pattern more reminiscent

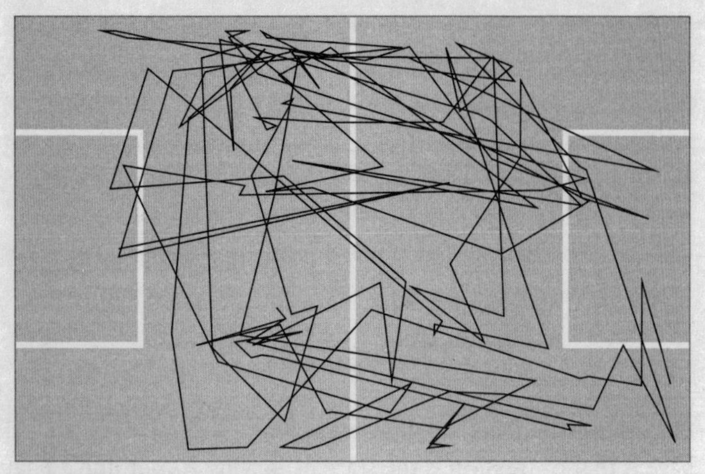

The line indicates the path of the ball

Figure 26 Ball movement between 11th and 20th minutes, Aston Villa vs Wolves, 19 March 2011

of a Jackson Pollock painting than a series of intentional ball movements.

At first glance, the ball's movement appears entirely random, its *x–y* locations on the pitch seemingly devoid of rhyme or reason. When we fill in the graph with data from the entire match, the lines become more numerous, but the pattern no more clear. It paints a picture of a game where the ball has a mind of its own, eluding any form of control or possession. Football's flow seems ever present.

That doesn't mean there is no point in players honing their skills in touching the ball with every permitted body part to try to influence its movement, its speed and direction. They might even generate something on the pitch that creates the illusion they possess the ball, if only because it is out of reach of the other side. But an illusion it is: no team has *complete* control of the ball, *except* when it lies in the goalkeeper's hands, or when they have a set piece. Only then are they truly in possession of the ball because the rules of the game allow them to be.

That has not stopped 'ball possession' becoming a cornerstone of our understanding of the game. Perhaps this is to do with football's close kinship with rugby and its cousin, American football, games in which discussing possession makes rather more sense.

But aside from set pieces, throw-ins and the safe hands of the goalkeeper, for the vast majority of the game a team does not have possession of the ball. It simply has *more* control over it, at that fleeting moment, than its opponents.

What matters in football, of course, is where the ball ends up: ideally, at the back of the other side's net. Teams are worried about what they can do to get it there and what the other side can do to get it into theirs. Possession, as we have seen, is something of a misnomer; instead, to understand the game better, we need to discuss how the ball moves around the pitch with more or less control by one side or the other.

Perhaps the most straightforward way to do that – to measure the various states of incomplete control that feet can have over a ball, and to understand how the ball comes to zoom about as in the game between Villa and Wolves – is to count how much and how often players touch the ball, moving it in their preferred direction.

Touching the Ball

According to data from Opta, over the course of a single Premier League season, all players together touch the ball about half a million times, give or take. That's about 1,300 times in the average match – 650 per team, or a little under 60 per player per match. They key word is 'touch'. To see how much more touching than actually possessing the ball there is in football, let us tell you about a clever little study.

Chris Carling, an English sports scientist who lives and works in France, has one of the best jobs in football. He is performance analyst for Lille OSC, the 2011 Ligue 1 champions. One of his chief concerns is how best to manage players' work-rates and levels of fatigue, both during a match and over the course of a long season.

For several years Carling has been investigating what are termed the physical activity profiles of professional footballers: measuring what it is that football players do on the pitch, for how long, how fast and to what effect. In one study Carling was interested in measuring precisely how much time individual players actually spend with the ball, how much running they do with it, and at what speeds. Using a multi-camera tracking system, Carling collected data from thirty Ligue 1 matches that mapped the movements of each player on the pitch.

Carling found that the vast majority of what players do doesn't actually involve the ball at all. And when we say 'vast majority', we mean it. When he isolated how often and for how long players actually touched or were in possession of the ball, the numbers were surprisingly low: on average, players had the ball for a total of 53.4 seconds and ran 191 metres with it during the course of a match.

To put these numbers in perspective, the time – less than a minute – that the average player spent with the ball made up only about 1 per cent of the time he spent on the pitch. The numbers are also striking if you consider that the total distance covered by the average player in a match is around eleven kilometres – so running with the ball made up about 1.5 per cent of the total distance each one covered.[5]

When players did have the ball, the average number of touches per possession was two, and the duration of each possession was a mere 1.1 seconds.[6] While the amount of possession Carling recorded varied by position, the critical part of the story is that players did very little that actually involved the ball – 99 per cent of the time they didn't touch it, and 98.5 per cent of the time they ran without it. When they eventually did touch the ball, it was gone in an instant.[7]

Carling's study is important for understanding what happens to the ball on the pitch. It demonstrates how little football players actually play, if by 'football' we mean running with or touching the ball. If, however, we consider 'football' to be very short individual possessions with frequent but only fleeting touches to try to move it to a teammate or away from the other side, then there is lots of football. This suggests that football is not about having the ball so much as it is about managing what seem like a succession of inevitable turnovers.

This means what we call 'possession' in football consists of two things: first, to touch the ball, and second, to keep touching it. And when it comes to the latter, it's a question of *how much* and a question of *how well*. That means there are two qualities to possession: how many times a team gets the opportunity to move the ball, and the length of time teams end up having the opportunity to move the ball.

These are not the same thing. Theoretically, having more opportunities to touch the ball doesn't have to be a good thing. Surely a team's ultimate dream would be to have just one opportunity to move the ball, straight from kick-off, and would then keep it for the rest of the half and score in the final second.

That is unrealistic. Practically speaking, to play possession football, we need our team to lose the ball less often, and to keep it away from their opponent for longer spells.

Possession Is Plural: How to Pass with No Feet

How many 'possessions', then, do football teams actually get? More accurately, how often does the ball change hands (or rather, feet) between sides in the course of a match? And what do players do with the ball on those few occasions they actually have contact with it?

The most straightforward way to calculate the number of possessions is to add up the number of times a team loses the ball to an opponent during the match. In the average American football game, each team averages about 11.5 spells of possession, with the number usually between 10 and 13.[8] This means teams turn the ball over about 23 times per NFL game, and between them they have 23 opportunities to do something positive with the ball (of course, they get several tries per possession).

In basketball, the sport of abundant shooting and scoring, the number of possessions and turnovers is much higher – about ten times as high. In the typical NBA season, teams average between 91 and 100 possessions per game: a total of between 180 and 200 for both sides combined.[9]

And football? First we need to find a way to calculate what constitutes a single possession in football. Let's consider the high end of control: those times when a player wins the ball and the team then makes at least two consecutive passes or takes a shot. Opta Sports collects such a stat to denote teams winning controlled possession, though their term for them is 'recoveries'. Over the past three seasons of the Premier League, Opta's data show that teams gained possession in this way about 100 times in a typical match, for a match total of around 200. So on the conservative end, teams have at least 100 possessions of more than just a transient touch of the ball – a number similar to that of basketball teams.

If we define possession changes more loosely, though, and include all the times when the ball changes from one team to the other, giving one team the chance to create something, the picture changes considerably and football looks even more inefficient – a game closer to ping pong than to basketball. Including all those instances when the ball is intercepted, a player is tackled and loses the ball, fouls are conceded, shots go off target or the ball is passed straight to an opponent, the number of turnovers almost doubles. In the past three years Premier League sides have turned the ball over about 190 times per match, producing a total of 380 turnovers per game.[10]

In the average Premier League match, 10 of the 100 strictly defined possessions yield a shot on goal and only 1.3 in 100 possessions yield a goal. If we use the looser definition of turnovers and possessions, 6 in 100 loose 'possessions' yield a shot on goal, and 0.74 in 100 of these actually yield a goal.[11] Football is not a possession sport. It is a game of managing constant turnovers.

This holds true even at the elite level, and even for those teams who pride themselves on managing possession, like Arsenal. According to Opta, in three seasons, Arsène Wenger's team never had fewer than 140 turnovers and sometimes they had as many as 240, for an average of 175.

In fact there's relatively little difference across clubs, regardless of whether they have a philosophy of playing 'possession football'. Over three seasons the top ten clubs in the Premier

League allowed their opponents 101.4 strict possessions and 187.9 loose ones per match, while the clubs ranked eleven to twenty gave up an essentially identical 99.1 and 189.3 possessions. So possession isn't singular – in football, it's plural.

The typical Premier League side has almost 200 fresh opportunities every game to do something with the ball. Most of the time, whoever has it tries to pass it. The single most common action players perform are passes in all shapes and sizes: short, long, with the head or the foot, crosses, goal kicks, flick-ons, lay-offs – passes account for well over 80 per cent of events on the pitch. The next largest categories of ball events, at 2 per cent or less each, are things like shots, goals, free kicks, dribbles and saves. Possession, boiled down, is delivering the ball to a teammate. Possession is turnover-free passing.

This also means that possession requires a collective, rather than individual, effort. It is a measure of team competence, not a specific player's brilliance. To see this more conclusively, we can look at data analysed by Jaeson Rosenfeld of StatDNA. Rosenfeld was interested in working out how much a player's pass completion percentage is determined by skill – something the player has control over – rather than the situation he finds himself in when making a pass. Rosenfeld's hunch was that pass completion percentage had less to do with the foot skill of passing the ball and more to do with the difficulty of a pass the player was attempting in the first place. It was not, he thought, so much what you did as where you were.

To test his intuition, Rosenfeld turned to the numbers: specifically, 100,000 passes from StatDNA's Brazilian Serie A data. To assess a player's passing skill, he had to adjust pass completion by the difficulty of the pass being attempted. Surely passes in the final third of the field and under defensive pressure were more difficult than passes between two central defenders with no opponent in sight.

Once he had taken into account things like pass distance, defensive pressure, where on the field the pass was attempted, in what direction (forward or not), and how (in the air, by head, and one touch), a curious result emerged: 'after adjusting for difficulty, pass completion percentage is nearly equal among all players and teams. Said another way, the skill in executing a pass is almost equal across all players and teams, as pass difficulty and pass completion percentage is nearly completely correlated.'[12]

Think about what this means. It is virtually impossible to differentiate among players' passing skills when it comes to executing any given pass (at least at the level of play in the Brazilian top flight). Everyone can complete a pass and avoid a turnover in an advantageous position on the pitch if they are without pressure or playing the ball over only a short distance. As a result, at the elite level, the particular situation the passer finds himself in determines a player's completion percentage, not his foot skills.

While their *passing* skills may be highly similar, this doesn't mean that players have identical *possession* skills. The data do not describe what happens *before* the ball arrives. As Rosenfeld observes: 'Is Xavi an "excellent passer" because he can place a pass on a dime or is it more his ability to find pockets of space where no defensive pressure exists to receive the ball, with his ball control allowing him to continue to avoid pressure and hit higher value passes for an equal level of difficulty? Many players put themselves in difficult passing situations because they dwell on the ball too long and upon receiving the ball are not able to reposition their bodies in a way that opens up the field.'

Possession football, in other words, is more than just being able to pass the ball – at the very top of the professional football pyramid, it has relatively little to do with that: it is mostly about being in the right place to receive it, helping a teammate position himself in the right place in the right way, and helping him get rid of the ball in order to maintain control for the team. As countless coaches have yelled to many a struggling player, you don't pass with your feet, you pass with your eyes and your brain. Football is a game played with the head.

A good team, when further up the pitch, manages to create and find space for both the passer of the ball and his intended target, making the passing situation easier. A poor team, in the same place, would not create as much space, so the passing situation would be harder. Good teams are not better at passing than bad ones. They simply engineer more easy passes in better locations, and therefore limit their turnovers.

Passing the Ball: Quantity and Quality

Logically, the *number* of passes a team manages to produce in a match and a team's passing *skill* do not have to go hand-in-hand. A highly skilled team such as Internazionale, Real Madrid or Chelsea may choose to cede possession against, say, Barcelona because their game plan dictates that they should absorb pressure and play on the counter-attack. Conversely, a weaker team may play a succession of passes between unpressurized central defenders to wind down the clock, or to take the sting out of a game. How much you pass does not have to be the same as how good you are at passing.

Data from the real world, though, shows that possession is typically much more prosaic. As Figure 27 shows, passing skill and volume in the Premier League usually move in tandem. Teams that pass more often usually complete a greater proportion of them, and teams that complete passes at a higher rate get the chance to pass more often. Looking at figures from 380 Premier League matches – the entire 2010/11 season – tactics and skill in possession go together. Improving the odds that the passes a team plays find their man means more possession, over the course of both a game and a season.

Each circle in Figure 27 is a team's match performance. As pass completion percentage goes up, so does the number of passes per match a team accomplishes.[13] Averaged over the entire season, the picture of possession looks straightforward, as can be seen in Figure 28.

A team like Arsenal or Chelsea played more than 550 passes in the typical match; Blackburn or Stoke managed just over 300.

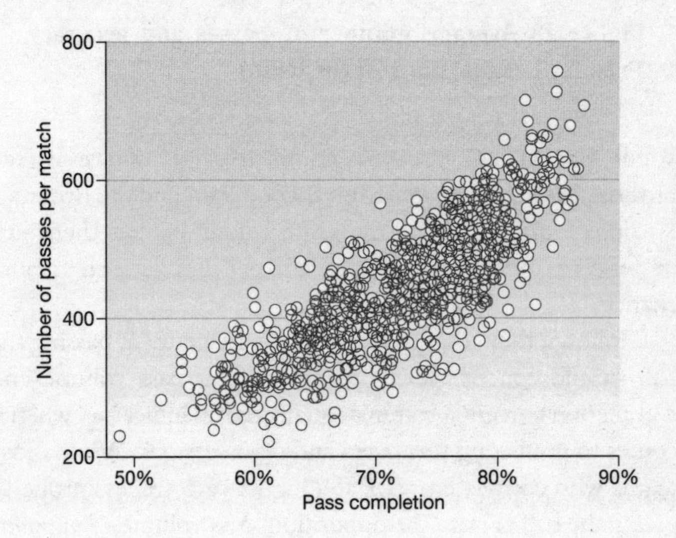

Figure 27 Average number of passes and accuracy, Premier League, 2010/11 (all matches)

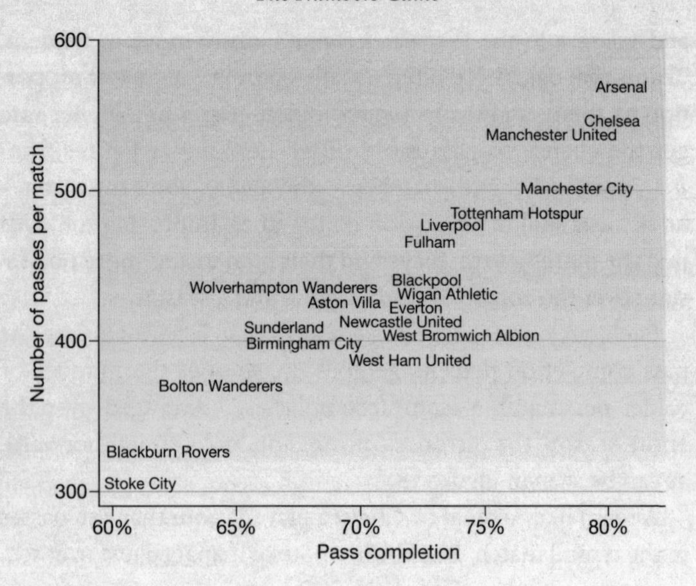

Figure 28 Average number of passes and accuracy, Premier League, 2010/11 (by team)

While Arsenal or Chelsea may complete eight out of every ten of those passes, the teams from Ewood Park and the Britannia Stadium found a man in the same colour shirt as them just 60 per cent of the time; that's only 10 per cent better than pure chance.

It follows, then, that the teams who are better at passing the ball should concede fewer turnovers. But pass volume and completion percentage aren't equally useful indicators when it comes to predicting turnovers and repossessions. While those teams who complete passes at a higher rate are less prone to giving the ball back to the opposition, pass volume – *how many* times a team passes – is only tangentially related to how often the ball is turned over.

While none of the teams that pass around 500 times a game or more turn over the ball a lot (Arsenal, Chelsea and the Manchester clubs), all the others give it away with varying degrees of frequency – unrelated to how many passes they have played. So in the 2010/11 season Sunderland, Aston Villa, Newcastle and West Bromwich Albion passed roughly the same amount, on average about 400 times per match; but they had very different turnover rates – at about 170, 180, 190 and 200 per match.

The teams who don't concede turnovers, who don't give the ball back to the opposition as much, are the ones that know how to play piggy in the middle. They can pass more safely around their opponents. They are not necessarily the ones who pass the most. Volume of passing is a tactical decision. The rate at which passes find their man is the true gauge of possession quality, and that completion rate is less about the fine calibration of the passer's foot than about the shared coordination of passer and receiver to create simple connections in difficult locations.

The Value of Possession(s)

There is, broadly speaking, a philosophical tension within football. There are those who prefer to see the ball swept about the pitch in beautiful patterns, the game played by Barcelona and Arsenal and Spain, inflicting upon their opponents a death by a thousand cuts. And there are those, José Mourinho and Sam Allardyce and the rest, who prefer to see attacks carried out quickly, efficiently and devastatingly. The former is often associated with beauty and the latter with ruthlessness; but such terms are subjective judgments, distractions designed to make randomness easier to handle.

The successes of Barcelona and Spain, though, have given

the passing school the advantage, for now at least. Passing's in fashion at the start of the twenty-first century. Possession, the theory goes, helps you win games. Have more possession, win more games.

We are not concerned with theory. We are concerned with facts. We wanted to know whether keeping the ball better gives you a better chance of success. If possession matters, we should see it reflected in results on the pitch.

Football analysts who have looked into this have often based their conclusions on their analyses of data from international competitions.

Twenty-five years ago, Mike Hughes from the Centre for Performance Analysis at the University of Wales Institute in Cardiff made the case that possession matters by analysing matches from the 1986 World Cup.[14] Hughes and his co-authors wanted to see if successful teams played differently from unsuccessful ones. Armed with a coding sheet for categorizing different events on the pitch and styles of play, they compared teams that reached the semi-finals with those that were eliminated at the end of the first round.

Their findings strongly suggested that possession matters and that possession football is a viable strategy for success. Successful teams had significantly more touches of the ball per possession than unsuccessful teams; successful teams played a passing game through the middle in their own half and approached the other end of the pitch predominantly in the central areas of the field, while the unsuccessful teams played significantly more to the wings. Finally, unsuccessful teams lost possession of the ball significantly more at both ends of the pitch – they turned the ball over more.

A follow-up analysis by Hughes and his colleague Steve Churchill based on the 2001 Copa América confirmed that successful teams played a different kind of football from unsuccessful teams. Among other things, successful teams were able to keep the ball for longer and create shots after possessions which lasted more than twenty seconds with more frequency than unsuccessful teams. They also were significantly better at transporting the ball from one end of the pitch to the other and into prime shooting areas. The data showed that the ability to pass effectively – again, to make complex situations simple – was at the heart of these teams' success.[15]

And it wasn't just the South Americans who successfully kept the ball. In 2004, a team of scientists from the Research Institute for Sport and Exercise Sciences at Liverpool John Moores University collected detailed data from forty matches that involved successful and unsuccessful teams in the 2002 World Cup tournament.[16] They too found that successful teams had a higher number of long passing sequences and made more consecutive forward passes.

But international competitions may be special: chance plays a disproportionate role in such tournaments, while the knockout format means we are only working with a small sample size of matches. What if we look at a league season? Academics P. D. Jones, Nic James and Stephen Mellalieu did just that, analysing twenty-four matches from the 2001/02 Premier League campaign to compare successful and unsuccessful teams.[17] Did possession matter for the outcome of any given game? Did it matter more at different times, depending on the score at that instant?

Yes, no matter where or when you looked. Mind you, both successful and unsuccessful teams had longer durations of possession when they were losing matches compared to when winning. Teams that were ahead gave the ball away more, and those losing by a goal or two chased the game and thus saw more of the ball. The real difference between victory and defeat was that successful teams retained possession significantly *longer* than unsuccessful ones, whatever the score was at the time.

Possession is related to success, not because of specific strategies related to what the score in the game was, but because of teams' relative skill levels. Possession is about ability, and that ability is chiefly to create easy passing situations where others would be pressured and face narrow windows. And that means that, over the course of a season, those teams who cherish the ball – and know how to treat it – will win out.

A Game of Two Halves

Most of Bill Shankly's wry observations on football have passed into folklore. But there is one that – at first glance – seems a little misguided. Shankly once complained that the Ajax who had scored five goals on a misty night in Amsterdam, with the young Johan Cruyff heavily involved, was 'the most defensive team we have ever faced'.[18]

We doubt Cruyff would dispute that description. The young maestro would have understood that having the ball is both an offensive and defensive measure. As he explained after orchestrating a 2–0 win for Holland against England at Wembley without ever crossing the halfway line: 'Without the ball, you can't win.' He would later add: 'If we have the ball, they can't score!' This should mean that, by completing more of the passes they attempt, by ceding fewer turnovers and by having more opportunities to pass, teams not only score more goals and concede fewer, but also win more games.

To find out whether Cruyff's assertions were true, we

looked at 1,140 matches over three Premier League seasons. That is 2,280 team performances.[19] The answers, shown in Figures 29 and 30, were clear.

In attack, teams who do a better job of keeping the ball away from their opponents *do* have more shots and *do* score more goals. In defence, they restrict their opponents to fewer attempts and they concede fewer goals. They have more shots on goal and suffer fewer. This, naturally, has a significant impact on goal production and goal prevention: teams that pass the ball well outscore their opponents by 1.44 to 1.19 goals per game, and they outperform them by an almost identical margin defensively. The data also show that, whatever possession statistic you look at – overall, completion percentage,

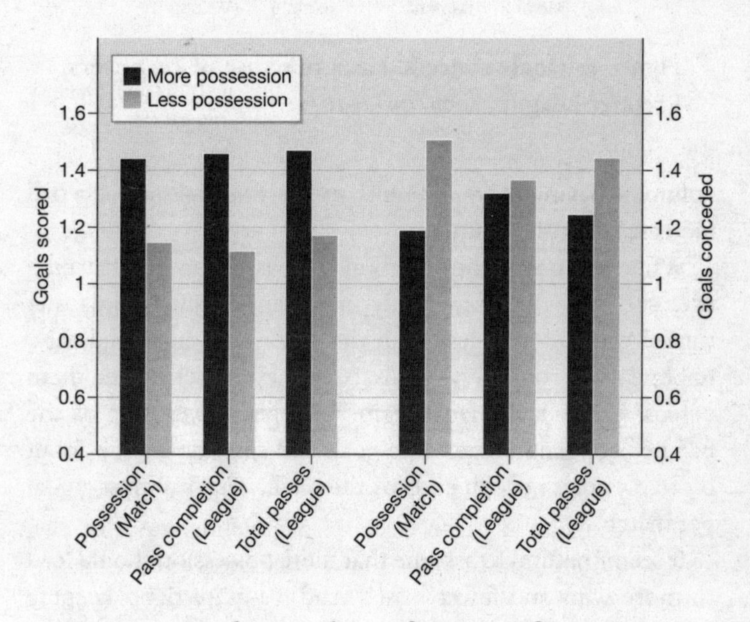

Figure 29 Goals scored as a function of possession, Premier League, 2008/09–2010/11

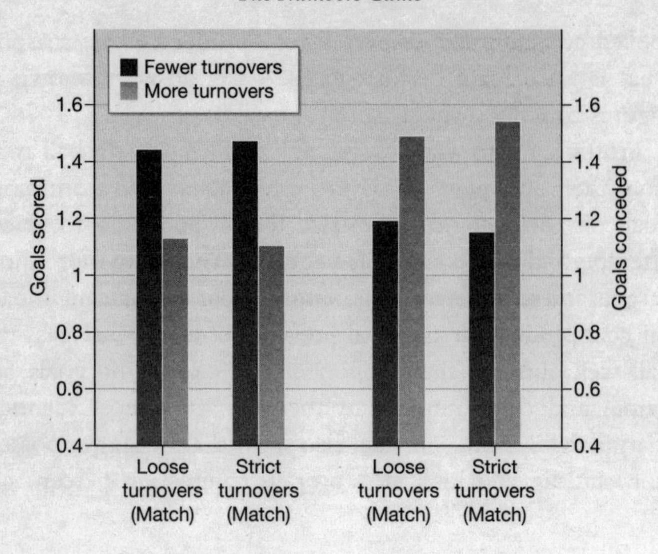

Figure 30 Goals conceded as a function of turnovers, Premier League, 2008/09–2010/11

volume – having more, rather than less, possession of the ball increases offensive output.

When we turn to the other kind of possession – not turning over the ball – we see equally important effects. Teams that turned the ball over less than the other side outscored their opponents by roughly 1.5 goals to 1.1; they outperformed them defensively by a similar margin.[20] Keeping possession of the ball helped teams score more goals and concede less by about 0.3 to 0.5 goals at both ends of the pitch. That's almost a goal per match.

It seems natural to assume that more possession should lead to more wins and fewer losses. And it's quite right: keeping hold of the ball, completing at a higher rate, and not surrendering it so often to the opposition means more wins, more

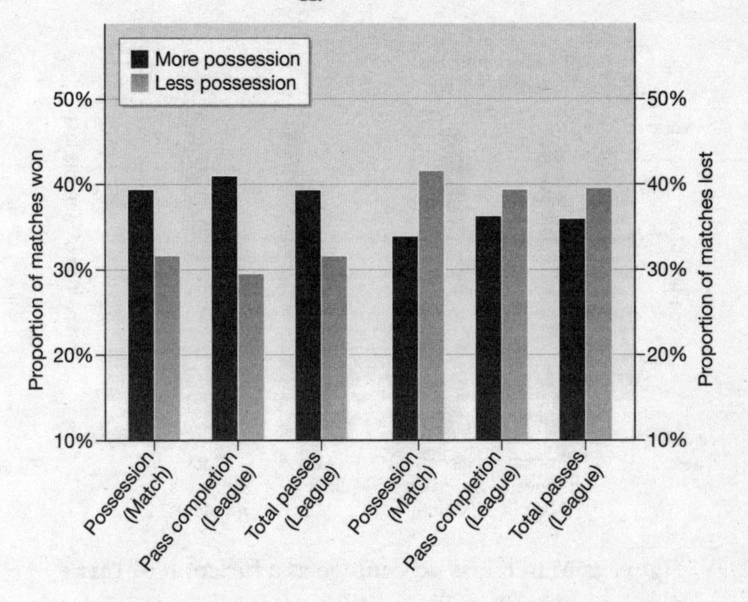

Figure 31 Match win percentage as a function of possession, Premier League, 2008/09–2010/11

points and more success. Teams that had the greater share of possession won 39.4 per cent of their games, compared to just 31.6 per cent if they had less. However possession is measured – volume, completion, or overall – having more of the ball generated between 7.7 per cent and 11.7 per cent more wins (Figure 31).

Pass completion percentages are nice, but avoiding turnovers is the most potent weapon of all. The teams that had less than half the turnovers in any given match won around 44 per cent of the time, while those that gave the ball away more won only slightly less than 27 per cent of the games. Having the ball is good. But not giving it back is better.

We have already discovered that titles are not decided only

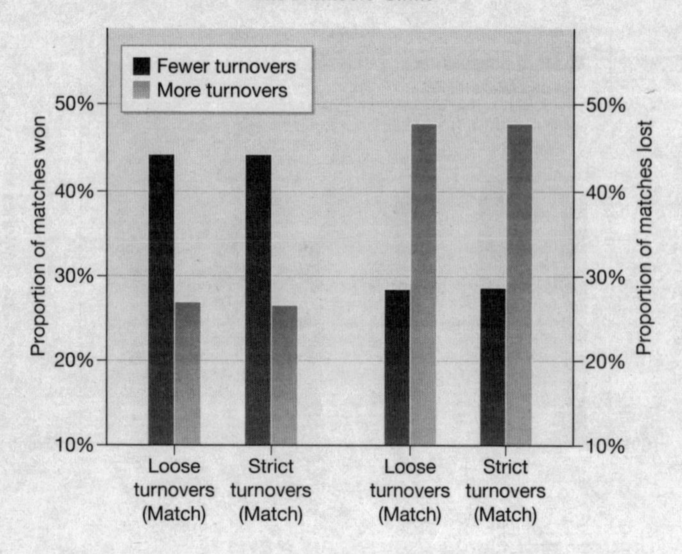

Figure 32 Match loss percentage as a function of turnovers, Premier League, 2008/09–2010/11

by winning: it is just as important not to lose. Possession helps here too. Having more of the ball decreased losses by around 7.6 per cent – about as much as it helped a team win. Turnovers, once again, are key: while pass completion percentage and total passes played matter much less for preventing losses than they do in garnering victories, not giving the ball away produced a staggering difference. Teams contrived to lose some 47.7 per cent of the matches in which they turned over the ball more than the other side, teams that gave it away less lost only 28.4 per cent of theirs (Figure 32). At both ends the possession game works, and with spectacular results.

All this pays off come the end of the season. Clubs that had more possession dominated the top end of the league and those that didn't were more likely to fight relegation. To see

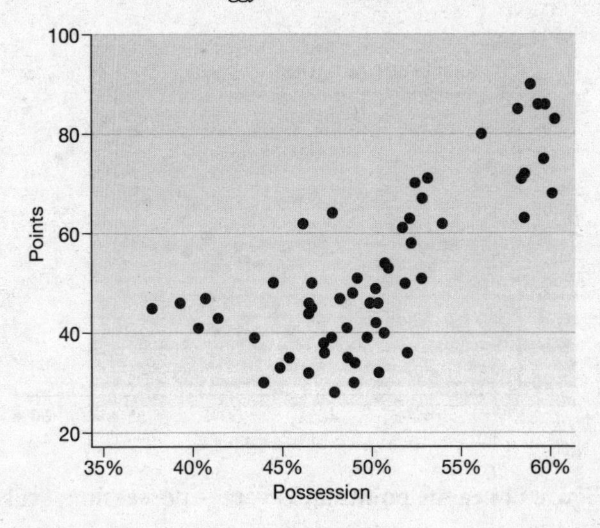

Figure 33 League points and average possession, Premier League, 2008/09–2010/11

how pronounced this pattern is, we plotted the number of points clubs produced in a season and the average amount of possession they had in the matches they played (Figure 33; each circle represents a club's performance for the year).[21]

Clubs with more possession will not win every match – far from it – but they will win more and lose less. The average league position of clubs with more possession than the opposition was 6.7; the average for clubs with less was 13.8. Ultimately more possession and fewer turnovers added up to a more successful campaign.

And yet, if we examine Figure 33 closely, we see that there are some distinct outliers to the overall pattern, especially on the left side of the graph. It seems there really are two distinct leagues in English football. In the bottom half are the teams with less possession, and in the top the teams with more.

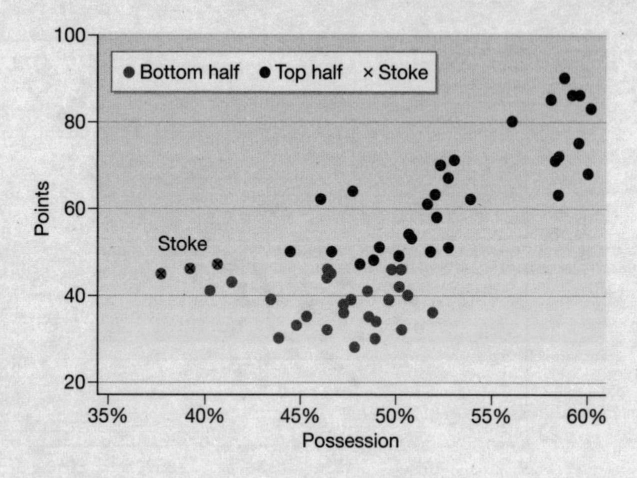

Figure 34 League points and average possession, Stoke City, 2008/09–2010/11

And if we look more closely still, we can see that, in that second league, there is one team that truly stands out. One side who win the battle for survival, over and over again, without seeing much of the ball. They even manage to finish above clubs with significantly more possession. That side is Stoke City (Figure 34). Somehow, Stoke have mastered the art of not having the ball.

Are they just a statistical anomaly, or do they have a secret?

6.

The Deflation of the Long Ball

It's not about the long ball or the short ball;
it's about the right ball.

Bob Paisley

Whether or not you like what Stoke do, it's hard to argue with their results. Until his sacking in 2013 manager Tony Pulis, an unremarkable defender during his playing days, had established the club not just as a Premier League side but as a cornerstone of English football thinking: how teams hoping to challenge for the title cope on a cold and windswept afternoon at the Britannia Stadium is often taken as a test of their credentials. New imports to the league are widely expected to wilt.

Pulis must take enormous credit. If Stoke were the Barcelona of route one football, then he was their Pep Guardiola. He made the *Financial Times* list of overachieving managers in England between 1973 and 2010,[1] and the authors of *Pay As You Play*, a path-breaking book on transfer finances, calculated that Pulis at Stoke spent less on transfer fees per point won than any other long-serving Premier League manager.[2]

But he has also faced enormous criticism. Stoke's long-ball style was considered unattractive and even philistine by many observers. Such scorn is borne out by the statistics: Stoke played more long balls and had less possession in the opposition half than any other Premier League team. According to these data Stoke should have disappeared long ago from the rarefied air of English football's top flight. And yet they continued to thrive. Why?

The answer was simple: Stoke were happy not to have the ball. In this age where possession is king, they were devout republicans. For Pulis, the Pep of the Potteries, less was more. It was as though Stoke believed they were more likely to score, and less likely to concede, if they didn't have the ball. And the only possession they really seemed to believe in was when Rory Delap was able to cradle the ball in both hands as he got ready to throw the ball into the box.

Stoke were perfectly happy to play less football than anyone else. Not just in the sense of not being concerned with getting the ball on the floor and keeping it, but in a very literal way. It was simple: the more the ball was in play, and the more Stoke had the ball, the worse they did. That is the key to understanding Pulis's success at Stoke and now at Crystal Palace.

When Less Football Means More Possession

In the course of one football match, nobody plays ninety minutes of football. According to Opta Sports, the ball was in play for between sixty and sixty-five minutes in a typical match across the four top European leagues in 2010/11. In the Premier League, the average was 62.39 minutes.[3] Yet for matches involving Stoke the average amount of time that the ball was in play that season was 58.52 minutes.

Stoke were like the schoolboy who takes the clock off the classroom wall, spins the hands forward, replaces it, and a few minutes later announces that the school day is somehow already over. In contrast, Manchester United offered the most action, with 66.58 minutes on average. Typically then, when the Potters were on the pitch, the ball was in play eight fewer minutes than when the Red Devils came to play. When we informed the head scout of another Premier League club of this and he passed it on to Pulis, the manager insisted he had no idea that was the case; this all just comes naturally to him, and by extension to his players.

He should not, though, have been entirely surprised. Under Pulis, Stoke systematically kept the ball away from the pitch. They were, in that sense, possession purists. They knew they only had real possession when their opponent put the ball out of play. Everything else was too uncertain. And so they maximized the one time that they controlled the ball absolutely: during set pieces.

That meant that the ball was in play significantly less in a Stoke game than those played by any other side. Indeed, this could be so extreme that in some Stoke matches, there was only around forty-five minutes of *actual* football. Stoke took a league-high 550 long throws in the 2010/11 season, and 522 the next year. Each time, Delap waited for the ball to be retrieved, gathered it in his hands, dried it with a towel, and the clock will have ticked. For those seconds, Stoke possessed the ball completely. They possessed it in a way that no other team could. The knock-on effect of this was to reduce their opponent's chances of getting the ball.

To an Arsenal fan like Rob Bateman, who lists Arsène Wenger as his sporting hero, this approach must seem abhorrent. Bateman, Content Director for Opta, regularly tweets facts such as this: 'Three of the four Premier League goals Stoke have scored against Arsenal have come from long throws. The other was a penalty.'

But that was not the only effect of Stoke's obsession with aerial bombardment. Their long throws created chances, but they also denied the opposition the opportunity to create their own.

This was the perfect strategy for Stoke because they were so bad at keeping the ball. According to analysis by Sarah Rudd, Vice President of Analytics for StatDNA, only slightly more than one in every ten possessions Stoke had in the 2011/12 season involved more than three passes. Only 4 per cent involved seven or more passes. This was football as Charles Reep envisioned it. Arsenal, by contrast, managed to produce four passes or more in 36 per cent of their possessions, with 18 per cent involving seven passes or more.

Or, more impressively: 43 per cent of the time Stoke had the ball, the subsequent movement stretched to precisely no passes. Almost half the time Pulis's team won possession, they gave it straight back. Arsenal, on the other hand, gave the ball away immediately just 27 per cent of the time.

Stoke seemed to understand that, for them, possession was actually counterproductive: the more traditional possession they had in a match – the more passing they attempted with their feet in open play – the more they lost the ball and turned it over to the other side, and so the more the other side had the chance to create opportunities. When Stoke had *less* possession than they typically had over the three seasons of data we looked at, they turned the ball over an average of 177 times in a match; but when they had *more* possession than usual, they lost the ball 199 times (a difference of 12 per cent). For a team like Arsenal, the exact opposite is true; Arsène Wenger's team turned the ball over 180 times when they had more possession than normal, but 186 times when they had less of the ball. The moral: when Pulis's Stoke had more of the ball, they lost it more frequently. When Arsène's Arsenal have the ball more, they lose it less.

These patterns of play have consequences for how Tony Pulis's men find ways to win. It's certainly much less from open play. In the Premier League overall, two of every three goals come from open play, and for high-possession clubs like Arsenal, it's as many as three out of four. In contrast, only half Stoke's goals resulted from open play. But they scored five times as many goals from long throw-ins as the average Premier League club. Another way to look at these numbers is that the average team in the Premier League scored 0.85 goals from open play per match and Arsenal a whopping 1.39. Stoke managed a measly 0.51, just 60 per cent of the average team's output.

Wenger, Menotti and Cruyff would be horrified at such figures. But it works: there can be no doubt about that. Stoke have been a Premier League mainstay since 2008. They did what Watford and Wimbledon did before them. They found a way to beat the big boys by using the tools they had at their disposal rather than imitating everyone else. They did not look like they had possession, but they were definitely in control. And now, Pulis's Crystal Palace is also ignoring the ball, but producing points nonetheless: from his appointment in November 2013 to mid-April 2014, the club's odds of relegation have shortened from 88 per cent to 1 per cent. His clubs understand that possession is not so much having the ball as not turning it over to the other side.

Cruyff would not like it, but he would understand.

The First, Failed Revolution

Stoke were one of the very few teams in the modern game that Charles Reep would appreciate. On the face of it, they did not share the modern obsession with possession, especially when the ball was on the pitch. Quite right too, Reep might think. His numbers, collected over thirty years with notepad, pencil and miner's helmet, showed that more than 90 per cent of possessions ended after three passes, or even fewer. Reep spent almost fifty years watching teams give the ball away, over and over and over again. No wonder he concluded that possession was a myth.

In fact, he probably would have found the idea that ball retention would become an aim in itself vaguely comical. Stoke's approach, constantly manoeuvring the ball into positions of maximum opportunity, would be just the ticket; as with Watford and Wimbledon in the 1980s and Egil Olsen's Norway in the 1990s, these are his ideas made flesh.

Sadly for Reep, Stoke may be the last of a dying breed. There are the sides coached by Sam Allardyce, cut from Reep's cloth, but to everybody else the long-ball game seems anachronistic. It has been widely discredited over the last two decades, since the heyday of Graham Taylor.

There is a simple reason for this. Reep had it wrong. As we showed in the previous chapter, keeping the ball – and not giving it back to the opposition – is a legitimate strategy for winning football matches and not losing them. It improves the number of goals you score and limits the number you concede. Of course, Pulis understands this basic truth: it is just that his way of counteracting it is the polar opposite to the response of most managers. Stoke kept the ball; they just did not keep it on the pitch.

From the off, Reep focused on understanding what it takes to win football matches. His premise was simple: if you could maximize goal-scoring opportunities, you would win more games. And to do that, he determined, teams simply needed to be more efficient. To Reep that meant scoring more goals with fewer possessions, fewer passes, fewer shots and fewer touches, not more. Only two of every nine goals came from a move involving more than three passes, and it took nine shots to produce one goal, say, while half of all goals come from possessions regained in or near the opponents' penalty area.

As soon as his numbers bore that out, it is no surprise that Reep reacted in the way he did: why were teams wasting their time with inefficient passing, when they could maximize their number of goal-scoring opportunities by moving the ball quickly into the opposition's penalty box, or by regaining possession high up the pitch?

If this is sounding familiar, that's because it is: Reep's conclusions, adopted by the likes of Stan Cullis at Wolves and, decades later, Taylor at Watford, were used as the philosophical keystones of the long-ball game. His findings even found their way into Charles Hughes's book *The Winning Formula*, though the author explicitly denied drawing on the Wing Commander's work. Hughes, a stalwart of the Football Association for years, was made its Director of Coaching in 1990: he became in a sense the high priest of non-possession football.

There was just one problem.

True, few observers found themselves falling in love with this efficient image of football – as Brian Clough memorably said, 'If God had meant football to be played in the air, he would have put grass in the sky' – but football is a results business; if it had worked, then the aesthetes would have been quieted by the trophy-clinching pragmatists.

No, the real problem with route-one football was that it was only intermittently successful.

Central to Reep's view of football was that sharp, smooth declining frequency of passes and the plunging odds of scoring from a move that involved more than three players exchanging possession. The vast majority of movements ended with, at most, one completed pass while 91.5 per cent of passing movements never reached a fourth player, as the declining bar chart in Figure 1 on page 18 showed. With every additional pass, an attacking side became less and less likely to score. Combine that with the numerical importance of pressing in the opponent's half – 30 per cent of all goals came from what we now know as 'final-third regains' – and you have the cornerstones of the long-ball game.

This analysis has been investigated since Reep's heyday. When Mike Hughes and Ian Franks, professors at the University of Wales Institute and the University of British Columbia, set out to re-examine Reep's work they found – using data from the 1990 and 1994 World Cups – that same steep decline in movements involving more and more passes, and they found a similar effect in goal scoring associated with passing movements of different lengths.[4] Initially, they agreed with Reep.

As they looked deeper, though, things began to change. The fact that most passing movements end quickly and that most goals are scored after a very small number of passes does not necessarily mean teams should try to live up to Reep's vision of efficiency, getting the ball into a scoring position with as few possessions as possible. That conclusion is too simplistic; in some ways that strategy was actually very ineffective. Why? Because the frequency of goals is not the same as the odds of a goal being scored.

To explain that, let's take a look at penalties.

In the Premier League, since 2009, some 65 per cent of goals have come from open play, while just 8 per cent have come from penalties. Open-play goals, in other words, are more than eight times as *frequent* as those from the spot. But then the *odds* of scoring from a shot in open play are 12 per cent, whereas from penalties, the chance is 77 per cent.

For a manager, then, what is the more effective strategy: building a team to score from open play, because that is how most goals are scored, or building a team to win penalties, because that is the most likely way of scoring? Do you go for frequency, or should you opt for favourable odds?

Penalties might be rarer, but they are also more profitable. Open-play goals are common, but less of a sure thing. It is this distinction in the statistics that Reep missed, and it is this distinction that goes a long way to explaining the failings of the long-ball game and the rise of an obsession with possession.

Like Reep, Hughes and Franks noticed a steep decline in passing movements as more players became involved. But they also found that the length of passing movements and the odds of scoring were connected. The longer the passing sequence, the better the odds of it being capped with a goal. Hughes and Franks concluded that teams with 'the skill to sustain long passing sequences have a better chance of scoring'. In fact, as the number of passes in a sequence goes up – as far as six passes – the odds of scoring go up, too.

The key factor is shots – their frequency and the rates at which they produce goals. Hughes and Franks found that shorter passing movements are related to effective shooting: for moves of four passes and below, conversion rates were higher than for five passes or more. Reep was right on that score. With a shorter move, a goal is scored one in every nine attempts; for longer sequences, that rises to a more profligate one goal for every fifteen shots a team takes.

In isolation, this finding would lead us to conclude that longer passing sequences gave defences a chance to set up, minimizing the element of surprise and dislocation of the defence by the attack. But that greater efficiency of converting shots from shorter passing moves doesn't equal more goals. Why?

Reep's numbers weren't wrong; unfortunately, he just didn't analyse them deeply enough.

What Hughes and Franks discovered was that longer passing sequences also produced significantly more shots on goal, thus increasing the total number of goals teams score. Reep was obsessed with converting shots more efficiently; longer passing moves do not cure that, but they make shots more frequent. There is a trade-off between opportunities and efficiency: longer passing sequences mean more shots for the attacking team, but they also mean lower rates of conversion of shots into goals.

Possession skill, Hughes and Franks discovered, is often the key difference between successful and unsuccessful teams: conversion rates between those sides that succeed and those that do not are about the same, but the successful teams produce a third more shots than the unsuccessful ones. It takes, on average, nine shots to score a goal. You will score more goals the more shots you have, and you will produce more shots when you don't give the ball away, either because you have the skill or because you have the strategy to play the possession game.

When we applied this to the Premier League, it rang true. To measure teams' devotion to the long ball, we calculated each club's ratio of long passes to short passes. The higher the ratio, the greater a club's percentage of long balls in the typical match. The results are shown in Figure 35, and you'll notice Stoke hanging way off to the far right. Teams that pass the ball more and that relied on a short passing game – defined by our

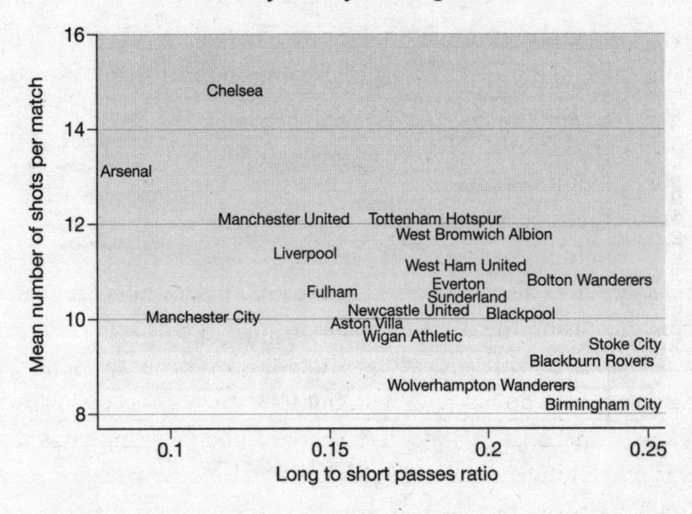

Figure 35 Long-ball ratio and number of shots, Premier League, 2010/11

research as any pass played under thirty-five yards – generated substantially more shots on goal.

That is the chief difference between success and – if not failure – then a lack of success in football. As may be seen in Figure 36, teams such as Arsenal, Chelsea and Manchester City – the sides that we found to play a possession-based game – had a similar conversion rate (goals from shots on target) to more direct sides; indeed, Stoke were actually more efficient in front of goal than Arsenal, whereas relegated Blackpool were roughly as effective as champions Manchester United. The difference is that Arsenal and Manchester United have 50 per cent more shots every game than those teams.[5]

The effect of this is clear: long-ball clubs have fewer chances to score and therefore score fewer goals, and they end their seasons battling relegation. Sides that treasure possession tend to be

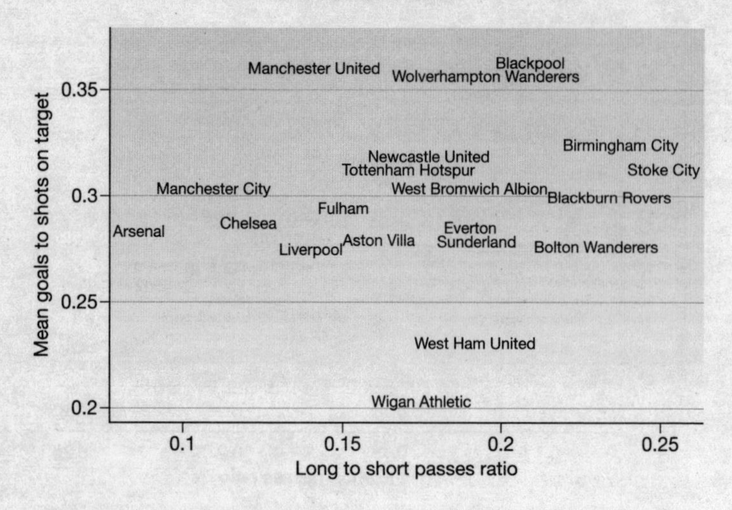

Figure 36 Long-ball ratio and conversion rates, Premier League, 2010/11

at the other end of the table, contesting titles (Figure 37). Those exceptions – from Pulis's clock-watching Stoke in Figure 37 back to Bolton under Sam Allardyce, who was among the first to apply analytics to the long-ball game – have found a style that helps them maximize their resources and fulfil their ambitions.

For them, the long ball is the right ball; they might never win the Premier League, but by perfecting their approach, at least both were able to secure their place in it for another season.

Restoring Reep

Football managers – never the fastest students in the class – appear finally to have worked this out. Reep's doctrine of maximal efficiency, the philosophy he and his followers had

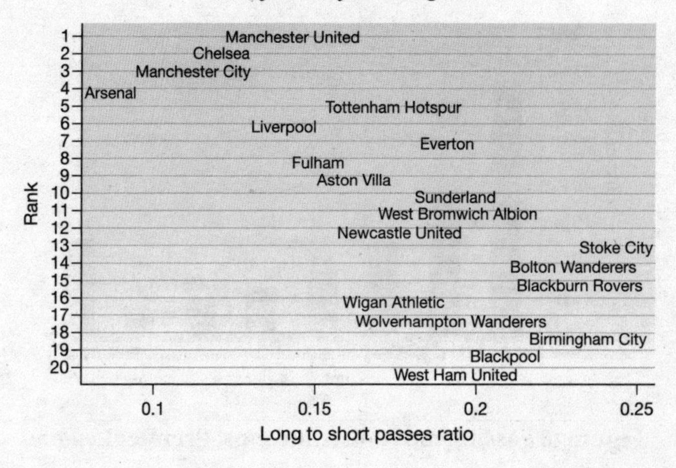

Figure 37 Long-ball ratio and league rank, Premier League, 2010/11

absolute faith in, is starting to disappear from the game. Yes, there are still teams who defy fashion – and logic – to play a more rudimentary, long-ball style, but the overall pattern is clear: possession, in the twenty-first century, is king.

That's what Sarah Rudd of StatDNA found when she looked at the passing sequences for the 2011/12 Premier League season (Figure 38). Reep's standout discovery, the curve that sharply declines with every extra pass in the move, has developed a spike in its tail. Advances in technology, training, technique and pitches have led to the dominion of the passing game. Moves involving seven passes are now as common as those composed of just two.

Yet it would be unfair to dismiss Reep as just a relic of times past. Yes, the football he espoused might seem a little dated, it may not be pretty to watch, and he failed to discover football's

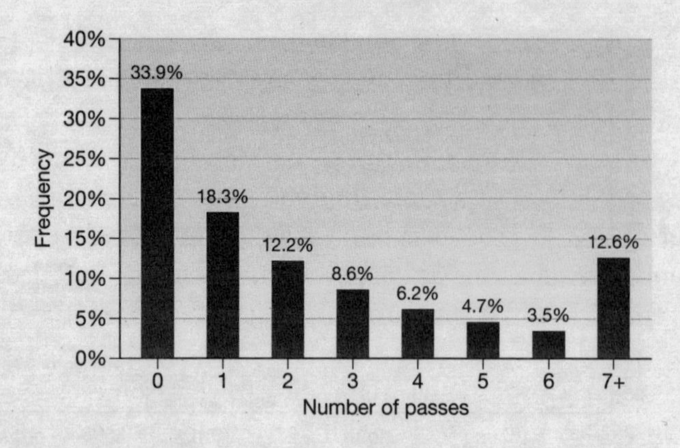

Figure 38 Passing move distributions, Premier League, 2011/12
Data source: StatDNA.

'winning formula', but his approach was in many ways thoroughly modern.

It was Reep who first tried to use data to help us see to the core of football, and in many ways it is from his work that the game's future will probably spring. He simply did not have the open mind or the techniques required to make sense of the wormball of information that every football match, every tournament, every season provides us. He recognized that football may look anarchic and disordered, but it can nevertheless be dissected into manageable elements, and those elements can be analysed.

We know the possession game is becoming more widespread, and we have the numbers to show that keeping the ball does help a team create more shots, and that more shots lead to more goals, and that more possession helps a side concede less frequently, which means they win more and, crucially, lose less. But is it the case that *every* team *must* play that way? No.

The very title of Charles Hughes's book was completely askew; Reep's aim for a universal cure for football's inefficiency was misguided.

There is no winning formula. But try telling Watford, Wimbledon or Stoke that the long-ball game doesn't work; try telling the Greece of 2004 that attacking football wins out more often than the defensive variety; try telling Barcelona or Spain to clear their lines. To each their own. As Bob Paisley, the Liverpool manager, once said: 'It's not about the long ball or the short ball, it's about the right ball.' For some teams, the long ball is the correct one. Indeed, as the possession game becomes ever more popular, the chances that there will always be one team playing in the style Reep preached increase. There will always be a benefit in going against the grain.

Reep was wrong on what the numbers implied; his findings were based on too rudimentary an analysis. But his assertion that football's numbers offered us a chance to see things that we had not yet glimpsed was absolutely correct. Unfortunately, Reep's system was peculiarly one-sided: it concentrated on how a team might best deploy its resources so as to score goals, rather than on how it might go about trying to keep them out. As we have seen, underestimating the role of the defence has been a characteristic of football ever since its first codification, and Reep was no different.

That is the failing of the long-ball game, too. It ultimately did not catch on as a generic prescription for a winning strategy, in part because it was too easy for more skilled sides to negate. It wasn't designed to adapt to a better opponent or teach a team how to keep a clean sheet. Ultimately, Reep wasn't a strategist and didn't know how to do defence.

There was nothing wrong with his general conclusion, though: it is in a team's best interests to be efficient. Bayern

Munich, in their Champions League final against Chelsea, or Barcelona, in the 2012 semi-finals, would both have welcomed the intervention of efficiency; for all their possession it was their profligacy that, ultimately, cost them the grandest of prizes. Efficiency was how Reep believed football teams could best overcome the role of fortune but he could never quite grasp the idea that his solution was not the only solution. There are many ways to control your own destiny in football. Perhaps the most effective way is not to be efficient; perhaps the most effective way is to control the ball.

It would be a shame to see Reep's legacy forgotten. Like many a revolutionary before him, he may have been a tad dogmatic, and a product of his time. But his was also the first sustained attempt at collecting football numbers and winning with them. The industry of data companies would not have evolved without him, and every club that has started out on its own journey to find out what the data say, owes Reep a debt of some sort.

Not every team wants to be Stoke. Not every team can be Barcelona. But every side can find a way to win, if they use all the intelligence at their disposal: that of their own talents and that offered to them by the numbers. That was at the heart of Reep's approach, and should not be forgotten. It is just that the numbers we have today are rather more advanced, rather more nuanced. Our intelligence – in terms of both gathering it and using it – is increasing.

7. Guerrilla Football

So it is said that if you know your enemies and know yourself,
you can win a hundred battles without a single loss.

Sun Tzu

No club in the Premier League generated less money than Wigan Athletic. No club in the Premier League had so little history, or so few fans. Ever since 2005, when they won promotion to the top flight for the first time in their existence, Wigan started the season listening to prophecies of doom. 2013 was the year that football gravity finally caught up with them, and they returned to their 'rightful' place among the also-rans. Even as the naysayers and doubters were ignoring seven years of wrong forecasts and congratulating themselves for seeing Wigan's fate, this little David took out one last Goliath, Manchester City, in the FA Cup final.

In their book *Why England Lose*, the football journalist Simon Kuper and the economist Stefan Szymanski found that money matters a great deal for the success of football clubs. According to their calculations, 92 per cent of the differences in English football clubs' league position can be explained by a club's relative wage bill.[1] It might not be the case that the team with the highest wage bill finishes top each and every season, but over the long term, the correlation is uncanny. At the other end of the table, it seems inevitable that, eventually, in football poverty will drag you down.[2]

For Wigan, this was unfortunate. The annual reports into football's finances prepared by the accountants Deloitte must have made miserable reading for anyone who followed the club: their turnover, wages and attendance were all fractions of the Premier League's giants. And yet Wigan managed to avoid relegation for seven years. It was almost pathological. They defied the laws of football economics. They disobeyed the laws of football gravity.

Part of the reason Wigan managed to survive so long in the rarefied air of the Premier League is Dave Whelan, the local magnate who owns the club. Wigan's average attendance was just 17,000 – they rarely sold out their home ground, the DW Stadium, its initials a (self-awarded) tribute to the club's benefactor – on a par with the likes of Vitesse Arnhem or the average German second-division side, but half the Premier League's average. That's a considerable shortfall in revenue. It's the same when we look at television and commercial earnings: in 2010/11, they earned £50.5 million from all of these streams – a tidy sum, to be sure, but half what the average Premier League team took. Only because of Whelan's enduring generosity did the club avoid sinking into the red. In 2011/12, he wrote off a £48 million loan to the club to balance the books. Financially, Wigan could not compete. And yet on the pitch they did.

In truth, Wigan did not dramatically outperform their wage bill, the gauge – for Kuper and Szymanski – of a manager's true impact. From 2006 to 2011, they finished eighteenth, fifteenth, fifteenth, sixteenth and sixteenth in the salary league, not far off their finishes in the actual division.

Yet Wigan's continued survival was still, as the respected financial blog The Swiss Ramble had it, 'a minor modern miracle'.[3] To explain why, we have to consider the odds that – given their spending on wages – Wigan would have been relegated well before the final axe fell in 2013. To do that properly, we need to calculate the odds of relegation as a function of a club's payroll.

The notional odds of relegation from the Premier League in any given season, for any team, are 15 per cent: three sides out of twenty endure the pain of demotion every year. But of course those three clubs are not simply drawn out of a hat: money does matter. More specifically, when we examined twenty years of club finances with the help of data from Deloitte, we found that a club's odds of relegation are 7.2 per cent if its wage spend is greater than average. In other words, you can halve the chances of being relegated just by spending a little more on your salaries than the average side. But for clubs that spend less, the odds of relegation shoot up from 15 to 21 per cent. For a team that spends as little as Wigan or less, these odds can even be as high as 44 per cent in any given season.

Spending less isn't a death sentence, but you are flirting with the chair. And spending less than the average year after year means the odds of relegation accumulate. For Wigan, the odds that they would be relegated at some point over the five Premier League seasons to 2012 were 95 per cent. It was, both mathematically and financially, almost a certainty. With wage bills four, two and one and a half times Wigan's £40 million, Manchester United, Aston Villa and Fulham faced odds of demotion of 0, 31 and 69 per cent, respectively.[4]

All this suggests that Wigan's continued survival was more than just good luck, and it was not simply attributable to their individual wage spending in any given year: the numbers were squarely against them. So Wigan's story is not just about money, but also how that money is put to use. There must be another factor at play. And we think it is that, rather than just using the story of David and Goliath as a clichéd parallel, they have actually learned their lesson from it. If you remember the tale, you will know that David could have taken Saul's armour and his helmet and tried to fight Goliath toe to toe. He didn't. He chose, instead, a very different stratagem.

Roberto Martínez: Insurgent Leader

By any standard measure Wigan had been a mediocre team for a long time. They conceded more goals than they scored in every season they were in the Premier League. They tended to have more possession than most of their peers at the wrong end of the table, but much of that came from the sterile domination of their own half.[5] Roberto Martínez's team, though, had been doing more than just passing the ball around at the back and getting lucky.

With the help of Ramzi Ben Said, a student at Cornell University, and the performance chalkboards published online by the British newspaper the *Guardian* in conjunction with Opta Sports, we tried to establish how Wigan went about scoring their goals in the 2010/11 season. Ramzi collected and coded a year's worth of data of attacking production (how each Premier League club scored their goals that season).

The data showed that the vast majority – 66 per cent – of the 1.4 goals a team scored in the average match that year came from open play. By far the smallest proportion of goals came from direct free kicks: just 2.8 per cent per team, per match. The average team produced one goal a game from open play, but needed to take thirty-five direct free kicks before finding the net that way.

But Martínez's Wigan were not your typical club. In 2010/11, they created goals in extremely unusual ways. They relied much less on traditional open-play goals than most, and did not bother with anything that resembled a patient build-up. In half their games they failed to score from open play at all. When they did, they tended to come from what are known among analysts as 'fast breaks' – lightning-quick counter-attacks.[6] And the rest of their goals came from free kicks. Their output in both these categories was exceptional. They scored twice as many goals on the break as the average side, and they scored almost four times as many goals from free kicks.

Rather than choosing one or the other, Martínez seemed to have forsaken both high frequency – not scoring from the most common source of goals – as well as good odds – trying to score from low probability shots (free kicks) – as a way to win matches. Martínez was not trying to fight his opponents in a conventional way. Instead, he was beating them any way he could.

Albert Larcada, an analyst at ESPN's Stats & Information Group, filled in the picture further. Using Opta's master file of play-by-play data, Larcada discovered Wigan were unusual in a number of other ways.

Not only did they score from fast breaks and free kicks, but when Larcada calculated the average distances from which Premier League clubs attempted shots that season, Wigan were the overall league leaders. Their average shooting distance was some twenty-six yards. This is why they were a significant outlier at the wrong end of Figure 36 comparing conversion of shots to goals and possession. This looked deliberate: their goals came from a longer distance than any of their peers – an average of 18.5 yards, way ahead of second-placed Tottenham, while Charles N'Zogbia and Hugo Rodallega both finished in the top five scorers from distance in the Premier League in 2010/11.

Martínez was thinking outside the box in the most literal fashion. Indeed, his team had the lowest number of goals scored from inside the penalty area of any side in the league – just twenty-eight, compared to Manchester United's sixty-nine.

This sounds very defensive – hitting teams on the break, relying on set pieces and long-range shots – but Wigan's formations told a more nuanced story. Opta's data showed that, while Premier League teams played 34 per cent of their matches with a traditional 4–4–2 formation that year, Wigan didn't play 4–4–2 in a single match. Instead their most common formation was a 4–3–3 system, usually thought to represent a more offensive tactical approach. Wigan's 4–3–3s accounted for one in eight instances of that formation in the Premier League. But they hadn't used it slavishly year in and year out. Instead, they adapted when necessary: Martínez masterminded his side's survival in 2012 by switching to a highly unorthodox 3–4–3 formation for the final third of the season.[7] It worked.

Martínez was trying to surprise his opponent and make sure he was not surprised himself. When we throw into the mix that Wigan led the league in recoveries, a clear approach crystallizes. Martínez's strategy relied on highly accurate long-range shooting, firing from distance – allowing his team to recover their defensive shape more easily – and persistence. He did not place any emphasis on corners – Wigan scored just one goal from a corner in the entire 2010/11 season – because it meant allowing his troops out of hiding and into open sight, leaving them vulnerable. Martínez was playing guerrilla football.

He had his team lie in wait for their opponents and then

punish them on the counter-attack. He employed sharpshooters, to let fly from distance, and snipers, to hit free kicks. His team were adaptable, unpredictable. With his neat jumpers and kind smile, Martínez looks a decent man. Underneath that veneer, though, beats the heart and mind of a natural insurgent, now headquartered at Goodison Park.

Intelligent Football

As it is for any revolutionary, information is at the heart of everything Martínez does. No rebel worth his salt would plan an uprising without gathering intelligence first, on the strength of his troops, of the ruling regime's weak points. The same principle applies in football.

That intelligence takes two forms: first, there is information. Managers have always gathered information in the traditional manner – scouting, talking to coaches, watching players in training, reading the news – and tapping into this network remains a crucial part of their work.

Most of that information, though, is subjective: to make the best decisions possible, managers must also tap into the objective sources of knowledge available to them. This is where the numbers come in. Nothing is more objective than data. Every manager now, whether he knows what to do with them or not, has one or more match analysts, housed at his club, with whom he will examine previous games and prepare for forthcoming battles.

Others are even more obsessive: Martínez, we suspect, is not the only manager to have his home TV connected to a data-analysis software package. Thanks to companies like Opta Sports, Amisco/Prozone, StatDNA, Match Analysis and all the others,

Martínez and his peers can now call up at the touch of a button accurate data on all their team's corners, or shots, or passes. Managers are inundated with numbers. Yet having facts at your disposal is not the same as knowing what each of them means.

The data collection companies are working on this. 'A lot of the innovation comes in figuring out what you actually need to measure,' Jaeson Rosenfeld, founder of StatDNA, told us. 'The problem is that you have to define a set of data that is complex enough to reflect what's going on in the game but simple enough that you can collect and analyse it.

'You could easily come up with a model that reflects a player's contribution, for example completed passes in the final third of the field. And you could come up with a hundred reasons for why that makes sense. But that's not enough. Several levels of detail deeper is what really matters. There is a lot of data out there already, but there is a high premium on insight.'

That is the issue for managers, like Martínez, as they contemplate how to plan their insurrections. They have all the knowledge they could possibly hope for, about their own teams and opponents. But which parts are important? This is where the second part of football intelligence comes in: deduction.

Football has been slow to accept analytics but gradually it is starting to infiltrate every corner of the game. Managers, and their employers, want an edge, that extra few per cent. It would border on professional negligence not to at least consider the numbers when so much is at stake.

Performance analysts are part of the fabric at most clubs. They are still not being utilized quite as fully as they could be, but the curve is upward: their influence can be felt in training, scouting and match planning. The next horizon, for John Coulson, Opta Sports' man tasked with managing the company's relationship with clubs, is in tactics.

'There is a strong resistance to statistics at the coalface of the game. Coaches have their jobs on the line and naturally rely on intuition and experience,' he told us. 'Clearly the role of stats is not to replace but complement these skills.

'However, with football being such a dynamic game and those at the forefront of it not being from an analytical background, it is challenging to build their trust in the metrics. Data is now readily available, and the next five to ten years will be about demonstrating the value deeper analysis of it can bring. We believe there will be another tipping point once someone proves by acting on it that there is a significant advantage in the data alone, much like we've seen in baseball and basketball.

'It has taken ten years to reach a tipping point whereupon video analysis solutions are widely adopted and used alongside data as evidence-based feedback to players and for opposition scouting. The message – that these programs are simply a tool to assist the coaching cycle – is now accepted but has been very difficult to get across. Inherently there was an opinion it can't show coaches anything they don't already know. However, the next progression – which is the use of advanced data analytics to actually influence tactical decisions and drive recruitment – is still very much in its infancy.'

The objection to the idea that the numbers can help is always the same: football is too fluid, too dynamic, too continuous to brook such a breakdown. That a problem has not been solved, though, does not mean it never will be. Yes, football is fluid, but that doesn't mean it can't be poured into various bottles. The possibilities are endless: open play versus dead ball; types of shots; penalty kicks; the timing of goals; tactical formations; home versus away; pitch location; what happens when teams are level, ahead, or behind. The race is on to find the best ways to break down the game and do it in a way that produces insights for ways of playing and valuing what players do. Moreover – as physicists and engineers who study interstellar nebulae, oil pipelines, or motorway traffic will testify – dynamic objects can in fact be analysed quite thoroughly.

There is a prerequisite before all this intelligence can be put to use: an appreciation of the simple but powerful fact that there is no 'best' way to play football. Scoring more is better than scoring fewer, and conceding fewer is better than conceding more. But beyond that, there is no easy answer.

Successful managers like Martínez understand this intuitively and they use the information available to them to craft a strategy that works for them in a particular moment. That could be the long-ball game, it could be a lightning-quick counter-attack, it could be trying to starve your opponent of possession. The guerrilla must adapt his tactics. As Gianluca Vialli and Gabriele Marcotti explain: 'Deconstruct tactics, and you find that basically it's a way to minimize a team's weaknesses while maximizing its strengths. That is what it boils down to. The concept is simple: it's about gaining an advantage over your opponent, and it has been around for thousands of years.'[8]

Tactics are not the same as strategy. Your strategy is what you plan to do over the entire season. Your tactics are what you do to get you there in the course of an individual game. To fulfil your strategy, you must get your tactics right; and your tactics must always fit your team *and* your opponent.

Going For It on Fourth Down

That analytics remains a source of suspicion for some is testament to the power of convention. There is a way that things are done – i.e. without analytics – and doing them differently, at least initially, is not tolerated. That is true off the pitch, in terms of how football has confronted the emergence of Big Data, and on it.

It is strange that two of the most competitive arenas in life – war and sport – should be dominated by so-called norms of behaviour. In an essay in *The New Yorker*, Malcolm Gladwell saw the same force at work in the story of David and Goliath.

'David initially put on a coat of mail and a brass helmet and girded himself with a sword,' Gladwell wrote. 'He prepared to wage a conventional battle of swords against Goliath. But then he stopped. "I cannot walk in these, for I am unused to it," he said . . . and picked up those five smooth stones. What happens when the underdogs likewise acknowledge their weakness and choose an unconventional strategy? When underdogs choose not to play by Goliath's rules, they win.'[9]

Gladwell argues this is true not just for battles for biblical supremacy but in any area of human competition where the weak face the strong. The best way for David to survive is to be innovative and do the unexpected. Their advantage, as Gladwell notes, 'is that they will do what is "socially horrifying" – they will challenge the conventions about how battles are supposed to be fought'. As importantly, to thrive, Davids have to work harder than the Goliaths. Wigan's insurgent football certainly fell into that category in 2010/11.

Though Martínez is one of the heroes of this book, he is far from unique. He is just the latest in a long line of clever managers who have found a way to unearth the value in his squad. These are the men who have changed the face of football for ever, by challenging prevailing wisdom and developing innovative approaches.

More often than not, these innovations were developed by teams that were not winning as much as they should, or simply were not winning at all. The strong do not need to innovate; it is the weak who must adapt or die. And it is to the managers of these weak teams that responsibility falls for finding the ways to innovate, to gain an advantage. If they fail to, it's their jobs that will be in peril.

It's these managers who have given us all of football's great innovations: the W-M – reportedly invented by Arsenal manager Herbert Chapman after losing 7–0 at Newcastle – *catenaccio*, zonal marking, the long-ball game. They are all attempts to upend convention and surprise the opposition. Knowing more, knowing better, knowing something new and knowing something different can help engineer wins or avert defeats. Aside from talent, hard work and swift feet, intelligence and innovation – on the pitch and off it – are key ingredients in success.

Playing the role of risk-tasking David is not foolproof, as Gladwell points out. 'The price that the outsider pays for being so heedless of custom is, of course, the disapproval of the insider.' The Goliaths are the ones who made the rules that insurgents challenge: 'And let's remember why Goliath made that rule: when the world has to play on Goliath's terms, Goliath wins.'[10] If David tries to beat Goliath at his own game, he will lose. He will not be criticized for failing in this manner; instead, his eulogy will be filled with patronizing praise. Imagine though if David's stones had missed their mark; his funeral would have been sparsely attended and his obituary intensely critical.

Playing unconventional football is an option available to anyone, not just the weaker teams. But the disapproval of the conventional world is hard to swallow. That is something perhaps best illustrated by dipping briefly into the world of the other football, the one so popular in America.

Kevin Kelley is the coach of the American football team at

the Pulaski Academy, a prep school in Little Rock, Arkansas – America's hinterland. He's enormously successful, but most of the American football establishment thinks he's got a screw loose. Kelley has worked out that some of the most conventional ways of playing American football lead to inferior outcomes – and yet virtually everyone in the game persists in them.

The most famous of these has to do with punting on fourth down. On every possession in American football, a team has four attempts (downs) to advance the ball down the field. If they gain ten yards, they retain possession for another four downs. If they have not managed to make ten yards after three tries, the team has to decide whether to try again or punt the ball far into their opponent's territory, ceding possession, but at least moving the danger further from their own end zone.

Conventional wisdom has it that it is better to punt the ball and keep the other side as far away from your goal line as possible rather than risk turning over the ball on fourth down. If close enough, teams will typically try to score a field goal, even though it is worth only three points while a touchdown is worth six.

In 2006 David Romer from the University of California at Berkeley wanted to know if it made sense to play the game that way. His research showed that punting the ball or scoring a field goal are actually the inferior choices – and yet this is what the majority of teams do on the majority of occasions.

Romer wasn't particularly interested in understanding football. Instead, he was concerned with discovering whether the traditional economic assumption that firms will maximize their options actually holds. His 2006 paper 'Do Firms Maximize? Evidence from Pro Football' showed that teams are consistently better off trying for a new set of downs on their fourth down attempts, and yet hardly any of them did so. Clearly, teams weren't maximizing their chances of scoring.

When Kelley, the Arkansas high school coach, heard about the study he felt vindicated and further emboldened. In his own football laboratory of a high school team, he had been experimenting with not punting for years and had been hugely successful playing what looked like a strange kind of football.

As David Whitley, a writer for the *Sporting News*, explained: 'At first people thought he'd lost his mind. "Idiot!" they'd yell when he went for it on fourth-and-eight from his own twenty-yard line. But the results justified the football heresy. Pulaski, which has only 350 students, has won two state championships. The current team is unbeaten and ranked No. 1 in Arkansas' Class 4A and No. 80 in the nation.'[11]

Defying convention has clearly worked for Kelley and his team. But when professional coaches play the numbers as Romer and Kelley see them, fans and pundits roundly criticize them. Perhaps the most famous example is the decision taken by the New England Patriots under their coach Bill Belichick to go for it on fourth down in the 2009 regular season game between the Patriots and the Indianapolis Colts, the closest thing the NFL had to a *clásico* during that decade.

Jeff Ma, a former blackjack player whose story was immortalized in the book *Bringing Down the House* and the movie *21*, backed that call:

> Belichick's Patriots were up by six points and faced a fourth down and two at their own twenty-eight yard line with just over two minutes left in the game. Rather than punt, which just about every other coach in the league would have done, he decided to go for it. Going for it on fourth and two at the twenty-eight yard line is successful 60 per cent of the time and,

> if successful, would effectively end the game. On average a punt from the twenty-eight would net thirty-eight yards. So a decision to punt would have to be based on an opinion that the extra thirty-eight yards was more valuable than the opportunity to end the game 60 per cent of the time.
>
> The advanced stats back up Belichick but I actually think this is a case where a seemingly counter-intuitive decision is actually very straightforward. Thirty-eight yards in field position is not worth giving up a 60 per cent opportunity to keep Peyton Manning [the Colts' excellent quarterback] on the sidelines.[12]

Sadly the Patriots didn't make that first down. They turned the ball over to the Colts, who drove down the short field to score a game-winning touchdown with thirteen seconds remaining. Belichick was ridiculed for not doing the 'right' thing. In reality, he had done exactly the right thing. It's just that it went wrong this time. But if you do the right thing often enough, the odds will be with you.

Know Yourself, Know Your Enemy

It is always hard to defend the unconventional in the face of defeat. Failure is accepted if you fail in a recognizable way. Nobody would have criticized Belichick if he had punted and the Colts had scored; just as a manager who employs man-to-man marking but sees his team concede at a set piece is not lampooned as mercilessly as one who uses the new zonal system. Doing the conventional thing may help a coach's job security; but the numbers can help him do the right thing and expand his ambitions beyond merely continuing employment.

The datafication of life has started to infiltrate football, and given managers, players, fans and observers the chance to see that the way things are 'always' done is not necessarily the way they should be done. Progressive managers understand that this new kind of intelligence is here to stay, so they will begin to make it part of their arsenal as they devise a game plan. Data can help you know your own team, and they can help you know your enemy.

We know there is no winning formula for football. Every team must change its approach every week, every game. Instead, a manager must know his players, his team, and he must know his opponents. He must make use of every resource at his disposal to gain every possible advantage. The numbers can help innovative managers hone their methods and accelerate the numbers game.

The idea that understanding your own team and the opposition is key is not new in football. Indeed, it explains an otherwise odd interest from many coaches in ancient Chinese philosophy.

Luiz Felipe Scolari – 'Big Phil' – and many others are devotees of Sun Tzu's *The Art of War*, a sixth-century BC treatise on military tactics. Before the 2002 World Cup Scolari gave a copy of the work to each of his players. It is unknown how closely Ronaldinho studied it, but his manager felt there was wisdom contained within, not least the quote which opened this chapter: 'So it is said that if you know your enemies and know yourself, you can win a hundred battles without a single loss.'

Managers desperate to win as many battles as possible will naturally turn to the insights held in the numbers; the trick is handling them correctly.

Take shots, for example. Knowing how many shots on goal the average team attempts is useful for giving us an overall indication of a team's offensive production. That pure number, though, does not tell us anything about the conditions under which those shots were produced or their quality, two things that can vary for many reasons, only some of which have something to do with players' skill levels. Finding those data requires a greater level of understanding.

Analytics can produce useful information about which actions on the pitch yield which results: do long balls create more chances than crosses? Does dribbling in your own half hurt your team, or the opposition? Is 4–4–2 a more effective formation than 4–3–3, and under what conditions against which opponents? They inform the way we play that game, how we understand ourselves, and how we approach our opponents.

What they cannot do is tell a manager how to go about implementing his strategy, or the tactics he needs to employ to get him there. It cannot tell him whether it is always better for his team and his players to try to keep the ball or whether it is always better to aim for a succession of lightning-fast counter-attacks or, as Roberto Martínez did with Wigan, instruct his team to play for direct free kicks and shoot from long range. The numbers contain a truth, not a set of instructions.

The data cannot do the manager's job. Numbers cannot put us all in the dugout; analytics is not an attempt to mechanize football. It simply enables him to do his job of building and directing a successful team with the clearest possible vision of what is happening on the pitch.

In the Dugout: Building Teams, Managing Clubs

8.

O! Why a Football Team Is Like the Space Shuttle

A team that commits errors in no more than 15 to 18 per cent of its actions is unbeatable.

Valeriy Lobanovskyi[1]

A battalion is made up of individuals, the least important of whom may chance to delay things or somehow make them go wrong.

Carl von Clausewitz

No single game in world football, the spiel runs, is worth as much money as the Championship play-off final at Wembley. Two teams from English football's second tier face each other in a winner-takes-all game for the final available slot in the Premier League. The total prize for the side that claims victory and promotion to the richest league in the world is worth around £90 million in television revenue, merchandising income and ticket sales. The new TV contracts signed by the Premier League may increase this figure to as much as £120 million.

The play-off final is not the sort of occasion in which you want to find your worst player is also your most significant.

Sadly that's just what happened on 30 May 2011, when

Reading and Swansea met for the right to claim a spot in the Premier League. This should have been a game in which one player's brilliance stole the show, making him a hero. Instead it was a game when Zurab Khizanishvili, Reading's Georgian central defender, turned into a villain. A football team is only as strong as its weakest link. And Khizanishvili, that day, was a very weak link indeed.

Everything that could go wrong, did go wrong. Chris Ryan, a writer for Grantland, was there that day, sitting among the increasingly exasperated Reading supporters. First, he wrote, they saw Khizanishvili get booked for fouling Fabio Borini, the Swansea striker. Then, in the twentieth minute, he bundled over Nathan Dyer, giving Scott Sinclair the chance to score from the penalty spot. Two minutes later, he failed to stop Dyer crossing for Sinclair to score his second.

That was not it. Before the first half had finished, the hapless Georgian had accidentally deflected yet another Dyer cross into the path of Stephen Dobbie, who duly gave Swansea what appeared to be an insurmountable lead.

'All around it was burst blood vessels, crying children and absolutely searing profanity and rage,' wrote Ryan of the Reading fans' reaction. 'It was basically several thousand people from Reading re-enacting the scene in *Goodfellas* when Ray Liotta finds out Lorraine Bracco just flushed all of his cocaine down the toilet. "Zurab! Why did you do that!?" Reading were 3–0 down after thirty-nine minutes.'[2]

Brian McDermott's team did all they could in the second half to rectify Khizanishvili's forty-five minutes of horror. They scored twice in quick succession and might have hauled Swansea back until the Welsh team scored a fourth and settled the game in the seventy-ninth minute.

Valeriy Lobanovskyi, the legendary coach of Dynamo Kiev,

would have been appalled by what he saw. To Lobanovskyi, a team's goal was to make mistakes in no more than 18 per cent of all of its players' actions. Reading, through Khizanishvili alone in a single half, had far surpassed that limit. As Jacob Steinberg summed up the first half in the *Guardian*, 'Reading have barely done a thing wrong, apart from pick one astonishing galoot at the heart of their defence'.[3] We do not mean to taunt the Georgian, or stir memories of what we suspect is the worst day of his career, but there can be little doubt that his mistakes may have cost Reading £90 million.

He is not the only man to have such a devastating effect on his team's hopes. Football is a team game, but it is one prone to being decided by sheer, staggering individual ineptitude. Every team has had one, a player whose very presence chills a fan's blood, whether it is William Prunier at Manchester United, Liverpool's Djimi Traore, Abel Xavier for Portugal, Jean-Alain Boumsong at Newcastle, Bayern Munich's Holger Badstuber, or even Marco Materazzi while at Inter: players who, with one misplaced pass, one lapse in concentration, can undo all the good work their managers and teammates have done over the course of a game, a week or, in Khizanishvili's case, an entire season.

Incompetence can also be communal. A team can be condemned by a lack of cohesion in defence, an absence of harmony and balance in midfield, or by apparent unfamiliarity in attack. All can doom a side's chances of winning a game or lifting a trophy. As Lobanovskyi recognized, football is a weakest-link game where success is determined by whichever team makes the fewest mistakes, whether they are individual or collective. The fewer Khizanishvilis a team has, the better its connections between its disparate parts, the better its chance of winning a match and the higher it will finish in the league.

This may seem obvious, but think of the consequence: if

football is a weakest-link game, where success is determined not just by what you do well but what you don't do badly, then it is by definition *not* a strongest-link game.[4] It is not the best players on the pitch or the strongest area of a team that decides who wins; teams who spend their summers lavishing millions on recruiting the latest superstar may have it all wrong. Football is vastly different from basketball, the most superstar-driven sport. It is less a result of Lionel Messi's majesty, of Paul Scholes's passing, of Cristiano Ronaldo's strength and speed, of Xavi and Iniesta's telepathic anticipation, than of the leaden boots and dull minds of Khizanishvili and his ilk, or their poor linkages with teammates.

If you want to build a team for success, you need to look less at your strongest links and more at your weakest ones. It is there that a team's destiny is determined, whether it will go down in history or be forever considered a failure. And that makes a football team really rather like a NASA space shuttle.

The Economics of O-Rings

Over the last quarter-century, as they have turned away from smooth supply and demand curves and idealized efficient markets and have started scrutinizing the rest of us, economists have begun to tell us an uncomfortable truth: it turns out that we, as a race, screw up all kinds of decisions. They know, for example, that we are unprofitably wedded to the status quo and are often controlled by the default options we face. In the United States, organ donation is rare because you have to tick a box on your driving licence if you wish *to* donate; in Europe, donation is common because you have to tick a box if you wish *not to*. We change our votes on important political referenda depending on whether lives are being saved or deaths are being prevented, two things that sound the same and are the same but are not treated as such. We are impulsive and impatient, we drink too much and we fail to save enough money for retirement.

The good news is that we do, at least, give economists a chance to create theories that take our imperfections into account. Michael Kremer, a very creative economist at Harvard University, invented one of the most influential of these 'flaw theories'.

Kremer's original article from 1993 was called 'The O-Ring Theory of Economic Development'. The name originates from the rings of high-tech rubber that were designed to seal the tiny gaps between the stages of the booster rockets that would lift the Space Shuttle Challenger into the sky in 1986. The rings, though, froze in the cold overnight temperatures at NASA's Kennedy Space Center at Cape Canaveral, Florida, and failed, allowing hot gases to escape and strike the enormous external fuel tank, eventually causing an explosion and the demolition of the entire vehicle, as well as the deaths of all seven crew on board. The failure of that one small part caused a sophisticated, complex, multimillion-dollar machine to malfunction. The O-ring was the weakest link in a system whose components and sub-processes were all integrated.

How can this be applied to economies? And, more importantly, what does it have to do with football? Kremer's theory can best be explained if we imagine an Economic League of Nations. Instead of points, table rankings are determined by a country's gross domestic product per capita – how wealthy a country is. The world is divided into three divisions: the US, the UK and most of western Europe, South Korea, Australia are in the top flight; relegated Russia, promoted China, India and Brazil among others are in the second tier; and finally

Honduras, Indonesia, much of Africa, central America and southern Asia are the third division 'clubs' with many poor supporters and low turnover.

In our Economic League, the following facts are true about the numbers in the three tables: wages and productivity are greater as you move up the divisions; there is a positive correlation in wages paid to different occupations (lawyers and bakers alike make more money in the UK than they do in Pakistan); rich countries specialize in complicated products; firms in wealthier countries are larger and invest in an 'efficiency wage' (spending time to ensure they recruit people who are suitable for the job and paying more money to induce loyalty and reduce turnover); and, lastly, firms will hire employees of similar skill and quality: to quote Kremer, 'McDonald's does not hire famous chefs, Charlie Parker and Dizzy Gillespie work together, and so do Donny and Marie Osmond.'

Kremer's insight was that many production processes – any time a group of people assemble to work together – are divided into 'a series of tasks, mistakes in any of which can dramatically reduce the product's value' or the overall success of the group's efforts.

One mistake, one slip, by one individual and the whole is affected.[5]

In general, workers execute a task with a certain efficacy. The most skilled worker may do a task at 100 per cent, while his less talented, motivated or knowledgeable co-workers make errors with varying frequency and scale, so that their individual quality on this task is 95 per cent, 82 per cent and so on. Sometimes in life, these errors add up but they won't cause a catastrophe. But in the kind of production process Kremer is worried about, the errors *multiply* rather than add up; the result, therefore, can be fatal. So, when the O-ring on the Challenger failed to perform its task, it zeroed out the entire shuttle.

How does this affect football? Think of a team as a small company with eleven workers, ten of whom are performing an equally important task at optimal, 100 per cent efficacy while the eleventh is only performing at 45 per cent capacity. In some economic processes the value of the final product is still 95 per cent (add all the qualities and divide by eleven), so the effect is minimal. But for an O-ring process, the value is 45 per cent (achieved by multiplying the qualities), and the product will be consigned to the bargain bin, the company declared bankrupt, the painting torn from the wall – or the team relegated.

We need to determine, then, whether football is an O-ring process. Whether one inefficient member, or a faulty connection between two players, or a rare mistake by a great player can significantly affect the entire team's performance. Does football have the hallmarks of those economies Kremer was discussing? Well, yes, we think it does.

Let's look at some numbers to see why. As in our imaginary Economic League of Nations, wages and productivity are much greater as you move through the leagues towards the top tier, as figures published by Deloitte show. As is evident from Figure 39, the ramp-up in wages between leagues is dramatic.

Deloitte's figures include the wages paid to groundskeepers and secretaries and other ancillary staff, but data unearthed by the website Sporting Intelligence relating solely to players paints a similar picture:[6] footballer wages in League One are twice wages in League Two; in the Championship, they are three times what they are in League One, and in the Premier League they are five times larger still.[7] It's pretty obvious why kids want to grow up and play in the Premier League, why Rolls-Royce salesmen don't bother bringing their business cards to Barnet FC's home games and why there is a Ferrari and Maserati dealership round the corner from Manchester City's training ground.

There is yet more evidence that football subscribes to all the criteria laid out by Kremer. While goals may not be the best measure of a team's productivity due to the heavy influence of chance, we can still use the number of shots, and shots on target, as reasonable measures (Figures 40 and 41). We would expect to see the same declining measure as we move down the ladder of English football in terms of these two factors as we do with wages. Just as wages are greater as you move up the divisions, so is productivity.

There is a positive wage correlation, too: just as lawyers and bakers earn more in the UK than in Pakistan, so do star strikers and their secretaries, coaches and publicists make more money at Manchester United than they do at Bradford City. In every

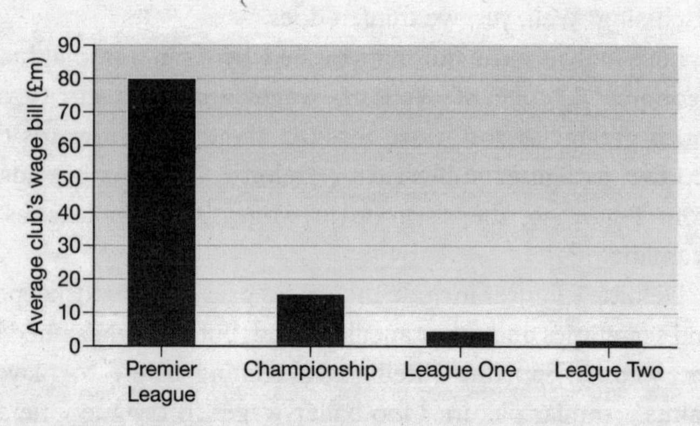

Figure 39 Annual wages in English football, 2010/11
Source: *Deloitte Annual Review of Football Finance, May 2012.*

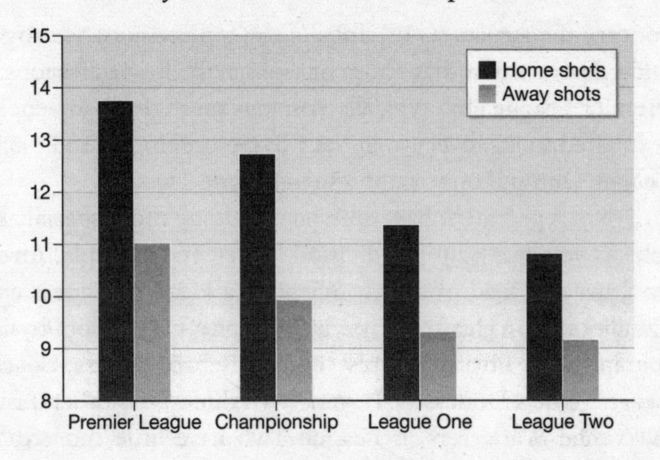

Figure 40 Average number of shots per team and match, 2010/11

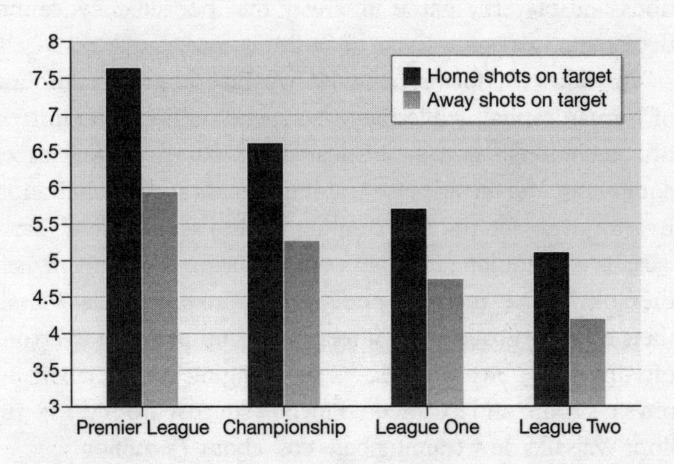

Figure 41 Average number of shots on target per team and match, 2010/11

country the structures of clubs in the top divisions are larger and more complex than those of clubs in the lower divisions: a Premier League club typically employs more than 350 people, compared to just over 150 in the Championship, around 100 in League One and only around 50 in League Two.

Every department has more people doing more specialized jobs as you move up the division ladder. For example, Liverpool have a Head of Sports Sciences, a Head of Fitness and Conditioning, a Head of Physical Therapies, two Senior Physiotherapists, a Physiotherapist and a Rehab Fitness Coach; League One's Doncaster Rovers have three Physiotherapists; Wycombe Wanderers in League Two have little more than three ice packs and a jumbo pack of sticking plasters.[8]

Just like rich countries that specialize in complicated products such as aircraft, software and luxury resorts, rich football clubs invest more capital and technology into their organizations and play the game in a way that poorer clubs cannot duplicate.

This takes two forms: richer clubs utilize far greater amounts of human capital, while they also spend millions on information technology and sophisticated databases, as well as on equipment and facilities for training, fitness and rehabilitation. Everton have ten full-size training pitches at their Finch Farm base, a well-equipped weights room, a state-of-the-art physiotherapist's suite, recovery pools and all the rest, while Walsall, their training ground just fifteen acres compared to Everton's fifty-five, have two pitches, some changing rooms, a gym, a physio's room and a canteen. Finch Farm cost around £17 million; Walsall's new training base cost about £1 million.

That complexity of product appears on the field. As the German journalist Raphael Honigstein noted in *Englischer Fussball*, his perceptive look at his adopted homeland's game, football is played in a much more sophisticated fashion in the Premier League – or the Bundesliga, or Serie A, or whatever – than it is in the lower divisions.

'At the very top,' Honigstein writes, 'pure Route One (that is, classic kick and rush) is generally proscribed and discredited as a tactic. One level down – below the radar, if you will – English football has preserved its own unique ideology: it's still a very territorial game. At this level, in other words, territory is often more important than the ball . . . Each corner is celebrated as if it were a last-minute winning goal. "Box 'em in!" the coach screams when the opposition have a throw-in near their own goal.'[9]

We have already seen that rich clubs pay their players more just as companies in richer countries do in Kremer's Economic League of Nations, but do they also spend more resources screening potential employees? There's no systematic information on the size of scouting networks – which operate on a relatively informal basis, with scouts, contacts and agents all recommending players – but there is abundant anecdotal evidence that this is an activity elite clubs invest much more time in than their counterparts in the lower leagues.

One hugely respected Premier League scout, someone as likely to be found on a Wednesday night watching the Champions League at the Nou Camp as he is to be at Harlington watching Queens Park Rangers' reserves, informed us in detail of the gulf between the top and the bottom in terms of the time and money they invest in evaluation and recruitment of players. He confirmed that the number of scouts at the top, middle and bottom clubs and leagues varies widely, and is typically tied to financial and league status. His best estimate was that the top Premier League clubs have fifteen to twenty of their own employees working on various aspects of scouting, from watching matches to providing background research and so-called 'technical scouting' – evaluating players' statistical information. With greater resource constraints and more holes in their squads, mid-table Premier League clubs will have about ten to fifteen scouts. Top Championship clubs have five or six employees engaged in scouting activities, while further down the table, perhaps three or four are. Once we move into League One and League Two, a club's devotion of precious resources to scouting activities quickly dries up, with perhaps two or three employees in League One, and fewer than that in League Two.

'There isn't much difference between Leagues One and Two', he said, 'quality of player-wise and otherwise. They don't have full-time guys working on scouting. Typically, someone has to double up and do opposition scouting, video analysis, and scouting players, or some combination thereof. But the jump to the Championship is noticeable, and the jump is even bigger to the Premier League.'

This is true across all Europe's major leagues, though there are some clubs where that gap is even larger. Udinese have around fifty full-time scouting and video analysis staff all round the world, as well as a vast informal network of contacts. It's this resource that has enabled the anonymous club from Italy's misty north-east to unearth some of the world's brightest young talent and transform itself into a contender for a Champions League place.

Because clubs at the top in Italy, Germany, Spain, France and England spend more time making sure they recruit the right players, it's no surprise to find that – according to figures from the CIES Football Observatory in Switzerland – those teams tend to hold on to their players for rather longer than smaller sides. The typical player will stay with a top side 30 per cent longer than he will remain at a lower-ranking club. That translates to an additional year or so: a significant portion of a player's career.

This is reflected in the length of contract on offer at clubs with different ambitions: according to the Premier League scout we spoke with, 'clubs in the lower leagues tend to give one- to two-year contracts, clubs in the Championship two- to three-year contracts, and in the Premier League, it's two to four years.'

This reflects the financial realities of life among the minnows. 'The lower league clubs have less control and more financial worry,' said the scout. 'They don't want to be tied to long contracts. Clubs in the Premier League make huge investments, and they want to protect those investments. One way to do that is to try and recoup that investment in the transfer market if things don't work out twelve to eighteen months into the contract. The last thing you want is for that player to be a free agent. In the lower leagues, it's too risky for clubs to give players long contracts. In the Premier League it's risky not to.'

Clubs hire players of similar skill and quality. Real Madrid will not bring in a journeyman midfield player from League Two – though they did their best when they signed Thomas Gravesen – while Alcorcón, the village team which knocked Real out of the Copa del Rey in 2009, will not go out and sign a superstar. This even has a fancy name in football's blossoming library of theoretical literature: the Zidane Clustering Theorem.[10]

In Defence of the Galácticos

Florentino Pérez's *galáctico* era at Real Madrid, the one that brought Zinedine Zidane, Luís Figo, Roberto Carlos, Raúl, David Beckham and Ronaldo together at the Santiago Bernabéu, looked like the worst kind of vanity project, one where an overwhelmingly wealthy nobleman unites all the top stars of the day at his court simply to feed his own ego.

History has widely written off the *galáctico* experiment as a failure. This seems a little unfair. It certainly ended badly, thanks to Pérez's inability to stand by his managers, his impatience, and his refusal to realize that maybe artisans are as important as artists. But it brought Real a Champions League trophy, their ninth, as well as a Spanish league title in 2003. Pérez might not have realized his dream of establishing a Harlem Globetrotters-style dominion over the game, but his money was not entirely wasted.

It has always been assumed that Pérez's first reign at Real – he has since returned to the club to try and repeat the trick, this time buying Kaka, Xabi Alonso and Cristiano Ronaldo, as well as hiring José Mourinho as coach – was an exercise in reducing football to its basest level. His plan, it seemed, was to take things like managerial judgment, scouting and team-building out of the game and, instead, simply buy the best players in the world. If he did that, Real would win everything.

This argument is the most extreme example of what happens when you think of football as a strongest-link game. By clustering stars together, Pérez was assuming that his team's overall performance would be multiplied by their excellence alone and not affected by the less-than-heavenly bodies that might be needed to fill out the squad.

Remember our earlier example of the company with eleven workers? Pérez's thinking was that if as many as possible execute their tasks at or near 100 per cent quality the total effectiveness would increase. This is not without merit. By replacing Guti – a player operating at, say, 80 per cent – with Zidane, who worked at 100 per cent, then Real's results would tangibly improve. This is the transfer market boiled down.

Teams try to replace their players with better ones, in the hope they will reap the benefits. That is why weaker players are cast aside or replaced and why superstars are bought.

Pérez knew that he could not afford to buy eleven superstars all over the pitch – or more, as injuries and suspensions would always mean he would need a squad. He could, at best, manage half a dozen of the very best in the world. The rest would have to come from the youth team. This was the policy of *Cracks y Pavones*, of superstars like Zidane and homegrown hopefuls such as Francisco Pavón, with the emphasis very much on the superstars. They would cover up for the weaknesses of the youngsters, while at the same time helping them to improve.

Across football, there is abundant evidence that players of equal quality do tend to flock together. This can be seen in the Fifa-sponsored Castrol Edge rankings, which evaluate every player in the top leagues in England, Spain, Italy, Germany and France on a monthly basis.[11]

Ian Graham, now the Director of Research at Liverpool FC but formerly of Decision Technologies, the company that developed the analytical evaluative system that underlies the Castrol rankings, touts their best attribute: 'a data-driven player rating system tells you the average thing that a player did'.[12] This means that the ranking reflects consistent production rather than a single glorious header or amazing back-heel. It also means that we can rank all players of a given club from the best to the worst.

The Castrol numbers from the 2010/11 season allow us to do two things: first, we can compare each club's strongest player with its eleventh-ranked player on a chart.[13] And second, we can compare players from different teams. If, in the real world of top-level football, the O-ring theory does *not* apply and there is no clustering of good players with good players and mediocre ones with other mediocre ones, the points depicting players' performances should either be randomly scattered around the graph or be gathered on a largely horizontal line. This would show us that strong players play alongside weak ones, as well as those of modest ability.

What we actually see in Figure 42 is a pronounced clustering of players with similar qualities. Great players play with other great players. For example the points in the upper right-hand corner represent Barcelona's best player, Lionel Messi, and the club's eleventh-ranked player that season, the defender Maxwell. Maxwell, in turn, is much more talented than the best player for the French side Arles-Avignon, the midfielder Camel Meriem, and also, as shown in the point closest to the bottom left corner, its eleventh-ranked player, midfielder Gaël Ger-

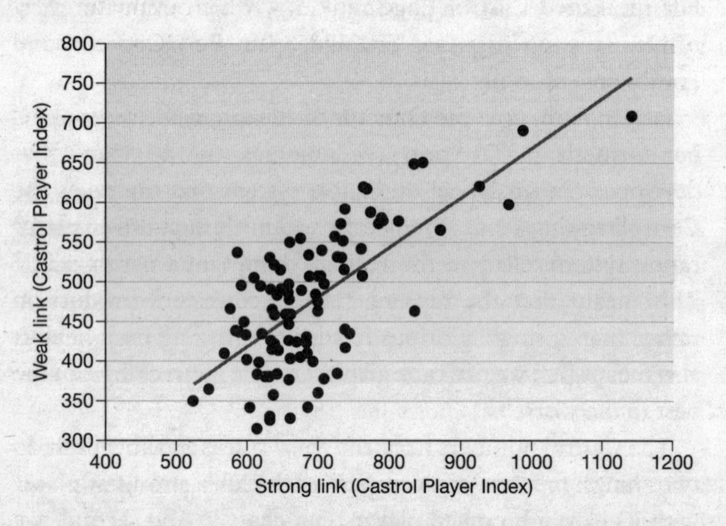

Figure 42 Top-ranked and eleventh-ranked European club players (Castrol player ranking points), 2010/11

Note: Goalkeepers are excluded from top position.

many. The correlation is so strong that it is roughly the same as the association between height and weight in the general population. Zidanes play with Zidanes.

More direct confirmation of football clubs being like space shuttles comes from some of the brightest minds in the game's history.

Take Arrigo Sacchi. Though not a top-level player himself, Sacchi was the mastermind behind the rise of AC Milan, making them the finest side in the world in the late 1980s. In 2004 the Italian was appointed as Technical Director of Real Madrid, brought in by Pérez in a bid to ensure the *galáctico* project remained on track. Sacchi was unimpressed.

'There was no project,' he said. 'It was about exploiting qualities. So, for example, we knew that Zidane, Raúl and Figo didn't track back, so we had to put a guy in front of the back four who would defend. But that's reactionary football. It doesn't multiply the players' qualities exponentially. Which actually is the point of tactics: *to achieve this multiplying effect on the players' abilities*.'[14]

The reason that talent does not always win out on the pitch is not just to do with the role of chance (though this is vastly important). It is because football offers so many ways to multiply your abilities, not simply add them. Tactics, for a start. A team of very good players who have had their skills maximized by the use of an intelligent tactic can beat a team of superstars whose talents are exploited, but not integrated. Sacchi understood this intuitively, and drilled it home to his own *galácticos* while at Milan – the Dutchmen Ruud Gullit and Marco van Basten – in a clever training drill.

'I convinced Gullit and van Basten by telling them that five organized players would beat ten disorganized ones,' he said. 'And I proved it to them. I took five players: Giovanni Galli in goal, Tassotti, Maldini, Costacurta and Baresi. The other team had ten players: Gullit, van Basten, Rijkaard, Virdis, Evani, Ancelotti, Colombo, Donadoni, Lantignotti and Mannari. They had fifteen minutes to score against my five players, the only rule was that if we won possession or they lost the ball, they had to start over from ten metres inside their own half. I did this all the time and they never scored. Not once.'[15]

Sacchi is not the only manager to have seen football this way. Valeriy Lobanovskyi, while propelling Dynamo Kiev to greatness over a period of more than thirty years, strived to multiply his own team's abilities, to make them more than the sum of their parts.

Lobanovskyi was a trained engineer, and a pioneer of the numbers game. Early in his coaching career he brought Dr Anatoliy Zelentsov to his side in order to collaborate on a scientific, systematic approach to football. Lobanovskyi had studied cybernetics, a field whose central concept is circularity, and which deals with problems of control and regulation in dynamic systems. He and Zelentsov viewed a football match as an interaction between two sub-systems of eleven elements (players), whose outcome depended upon which sub-system had fewer flaws and more effective integration. The key characteristic of a team is that 'the efficiency of the sub-system is *greater than the sum* of the efficiencies of the elements that comprise it'.[16] In another interview Zelentsov said, 'Every team has players which link "coalitions", every team has players which destroy them. The first are called to create on the field, the latter – to destroy the team actions of [the] opponent.'[17] Using different concepts, this describes an O-ring production process.

To these wise words we can add indicative statistics. Returning to the 2010/11 Castrol rankings, we can examine the connection between a team's weak and strong links and its goal difference and number of points earned. To do that properly, we had to transform the Castrol numbers into percentages.

Because the tasks of players differ by position, we gave each player a quality score based on the tasks of his position and relative to the top-ranked performer at his position. For example, in May 2011 Joe Hart of Manchester City was the top-ranked goalkeeper and so he will be given a score of 100 per cent, while all other clubs' keepers will be less than 100 per cent (their Castrol scores will be divided by Joe Hart's). The same applies in defence and midfield, though not in attack.

Forwards are necessarily different because of the numbers of the only true genius in the present football universe, Lionel Messi. Messi is to other forwards as Mozart was to Salieri, as Rembrandt was to the average court painter, and as Muhammad Ali was to Sonny Liston. Table 5 shows the percentage difference in the scores of the top-ranked and second-ranked players at each position at the end of the 2011 season.

Table 5 Percentage difference in the scores of the top-ranked and second-ranked players at each position at the end of the 2011 season

Position	Player Ranked 1st	Score	Player Ranked 2nd	Score	Difference
Goalkeeper	Joe Hart	792	Christian Abbiati	764	3.7%
Defender	Mats Hummels	872	Gerard Piqué	864	0.9%
Midfielder	Florent Malouda	834	Frank Lampard	820	1.7%
Forward	Lionel Messi	1141	Karim Benzema	987	15.6%

Lobanovskyi and Zelentsov would immediately diagnose that Messi's score arises from his inclusion in the Barcelona sub-system. (As several desultory performances testify, his effectiveness in the Argentine O-ring production process is greatly diminished.)[18] Here, Messi's score is so extraordinary that we have had to do what so many markers have failed to do and remove him from the game. Because he makes everyone else look so bad, we have had to use Real Madrid's Karim Benzema as a basis for all other forwards.[19]

Now we can take this relative quality index and redraw Figure 42, which showed the close association across European clubs between their strong and weak links. The result is shown in Figure 43. There are a number of clubs in Figure 43 lying relatively far from the trend line: these are clubs where there is a stronger or looser match between a club's weak and strong links. For both Barcelona and Real Madrid, the weak link is of higher quality than the strongest player for 80 per cent of the other clubs in Europe's top five leagues. Some teams are well below the trend line, because their pool of talent was relatively thin – Newcastle, Blackpool, Borussia Mönchengladbach – and there are some for whom the eleventh-ranked player was not that much weaker that the top player: Manchester City, Lorient, Hannover 96.

Nonetheless, the overall O-ring pattern still holds: Zidanes gather together in one dressing room and Khizanishvilis in another, usually a far damper and less well-appointed one.

Why Galácticos *Matter Less than Galoots*

For all Pérez's flaws he had seen a truth: because football is an O-ring process, good players do cluster together. But he had missed the ultimate conclusion of this idea: that it is the weak

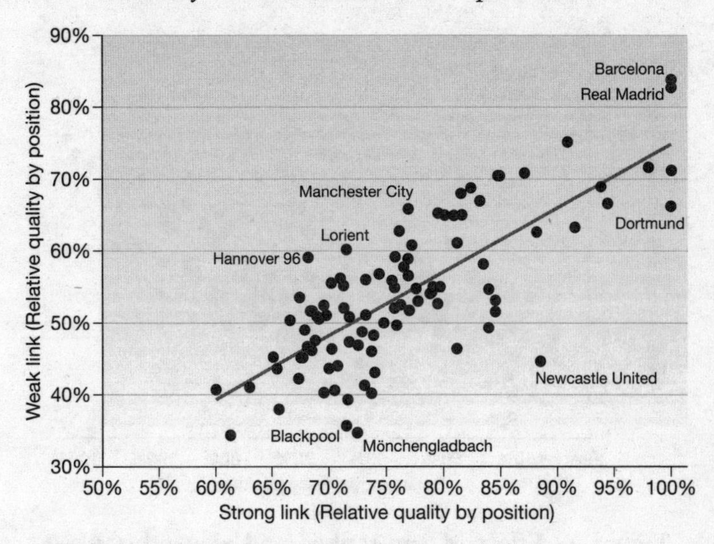

Figure 43 Top-ranked and eleventh-ranked European club players (relative quality by position index), 2010/11

Note: Goalkeepers are excluded from top position.

links who are the crucial determinant in a team or a company's success, not the strong ones.

To prove this, the critical test we need to conduct is to see how vital a role the weakest link plays in a team's success and ultimate position in the league table. Figures 44 and 45 reveal that the relative strengths of both the best and eleventh-best players are significantly and positively related to a club's goal differential for the season and the points secured in each game.

Every club on the chart has two points: Barcelona and Madrid are in the upper right, with Arles-Avignon in the bottom left. These are the clubs' strongest and weakest players. It's obvious that both are relevant to team performance. What

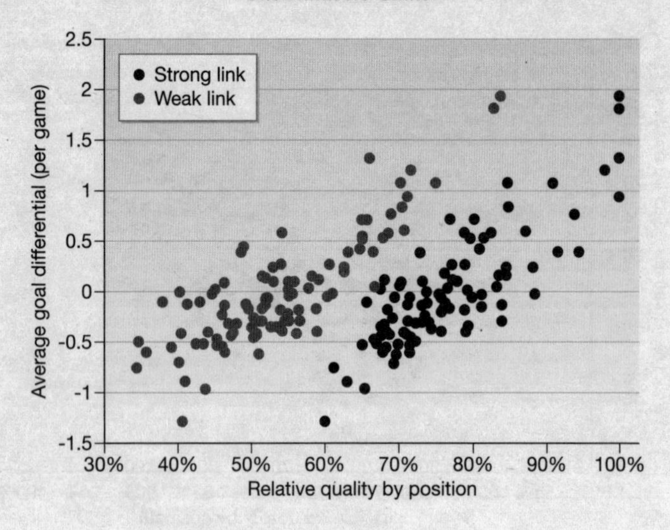

Figure 44 Effect of top-ranked and eleventh-ranked European club players on average goal differential, 2010/11

Note: Goalkeepers are excluded from top position.

isn't immediately apparent is which is more relevant: is football more a strongest-link game or a weakest-link one?

For this, we will need the most important piece of kit in the economist's toolbox: regression analysis. This will allow us to see if we can predict a club's success based on information about its weak and strong links, and which is a more powerful performance-enhancer.[20]

Once we apply these analyses – while accounting statistically for the differences across the leagues – we see that it is the *weak* link that matters more. For every percentage point that your best player improves, your goal difference per game increases by 0.027. That means that if you increase the quality of your best player from 82 per cent to 92 per cent, by signing a

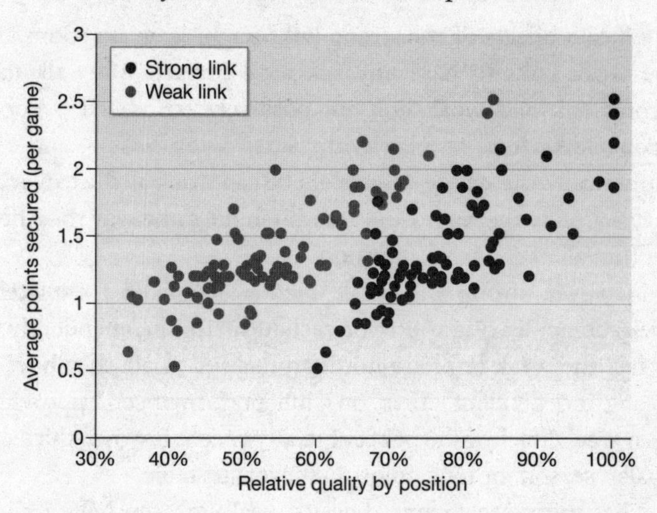

Figure 45 Effect of strongest and weakest players on average points won per game in top European clubs 2010/2011

Note: Goalkeepers are excluded from top position.

new striker, say, then over the course of a thirty-eight-game season, you will find your goal difference improving by just over ten. The results are just as demonstrable for points per game: the same upgrade on your star player would mean five more points a season.

For many teams, that is the difference between success and failure: a Champions League place compared to the ignominy of the Europa League, survival and relegation, winning the title or finishing an agonizing second. Those five points (for each 10 per cent upgrade) are why even very good teams are prepared to sink millions into another superstar signing.

Pérez felt that Madrid's superstars would compensate for any remaining weaknesses. Analytically, this could happen: the significant effects of the strong links might leave no room for the weak links to have any statistical impact. After all, the strong link and weak link are positively correlated – good strong links tend to play with better weak links – and so, numerically, the strong links might be capturing all the explanatory action in the regression analysis in the same way that they capture all of the fans' attention.

However, strong and weak links are far from completely overlapping, leaving room for each to matter independently.[21] In fact the weak links are not marginalized at all: they have a strong independent effect on club performance. Improving your weak link from 38 per cent to 48 per cent is worth thirteen goals a season, or nine points in the league table.

That means that upgrading its weak link can help a club *more* than improving its best player. Take the mid-table La Liga club Levante. With a strong link of midfielder Juanlu (quality 74.4 per cent) and a weak link of defender Juanfran (quality 56.8 per cent), Levante finished fourteenth in the league table with forty-five points in the 2011 season.

If through training, hard work, or magic Juanfran were to have boosted his quality by four points, then we would have expected the Granotes to jump up the table: they could have finished the league in eighth place with forty-nine points; in contrast, had they focused on improving their strongest link, Juanlu, by the same amount, their points total would have risen only by two and their table position by three spots.

The last way we can compare the importance of weak and strong links is to decrease or increase their quality by a commonly used statistical step, one standard deviation – a measure of the spread of qualities of all players around the mean. So, what happens to the average club if the form of its weak or strong link decreases a step due to, say, injury, or advances a step due to a transfer signing? Again, galoots are more influential than stars. The differences add up: a one-step decline in the form of your weakest link rather than your strongest link means 4.6 fewer points over the course of a season. More importantly, perhaps, improving your weakest link over your strongest link by one standard deviation translates into 13.7 more points in the final league table. Our results also show that performance differences in weak links are 30 per cent more important when it comes to goal difference, and almost twice as important with regard to points per game.

Imagine if Reading had not been forced to field Zurab Khizanishvili on that bright May afternoon at Wembley. Imagine if they had been able to choose someone just 5 per cent better. The whole course of football history might be different: maybe Brian McDermott would be Liverpool manager, not Brendan Rodgers, and Jem Karacan, not Joe Allen, would be at the heart of the Anfield midfield.

Or what if Pérez had paid as much attention to strengthening his *Pavones* as he had to garnering his *Cracks*? Perhaps then the *galáctico* experiment would not have, ultimately, disappointed. Perhaps he would have more than one Champions League and one La Liga title to show for all his hundreds of millions of euros in investment. He knew that football was an O-ring process. He just tried to solve it in the wrong way.

It is easy, as Pérez did, to think of football as a game of superstars. They provide the glamour, the genius, the moments of inspiration. They sell the shirts and fill the seats. But they do not decide who wins games and who wins championships. That honour falls to the incompetents at the heart of the defence or the miscommunicating clowns in midfield. Football is a weak-link game. Like the space shuttle, one small, malfunctioning part can cause a multimillion-pound disaster.

This has profound implications for how we see football, how clubs should be built and teams constructed, how sides should be run and substitutions made. It changes the very way we think about the game.

9.
How Do You Solve a Problem Like Megrelishvili?

The measure of success is not whether you have a tough problem to deal with, but whether it is the same problem you had last year.

John Foster Dulles

Some days are just bad days. Haim Megrelishvili, an unremarkable Israeli defender then playing for Vitesse Arnhem, probably did not know when he woke up on the morning of 15 March 2008 that he was in for a really bad day, the sort of day that would shock even Zurab Khizanishvili, our friend whose horror show cost Reading £90 million.

He probably did not even know it when he strode out at FC Twente's home ground that afternoon to warm up for Vitesse's Eredivisie game against the Enschede side. By the third minute of the game, when he allowed Romano Denneboom acres of space to receive a pass, failed to recover and watched helplessly as the striker fired the hosts into the lead, he must have realized he was not on top of his game. Still, it was probably a shock when, three minutes later, just six minutes into the game, he saw his number flashed up by the fourth official, standing on the touchline. Next to him was Alexander Büttner, a young left back who would go on to play for Manchester United, ready to come on. Before he had even broken a sweat, before all of the crowd were in their seats, Megrelishvili was being replaced.

Few players have ever been so publicly humiliated. It may not be the quickest substitution on record; it may not even be the quickest tactical substitution on record – examples from both Lincoln City and the Norwegian side Bryne beat Megrelishvili's moment of embarrassment for speed – but still, it was the sort of incident that haunts a player's sleepless nights. And it was, no doubt, compounded when his manager, Aad de Mos, did it again and withdrew him in the fifteenth minute of Vitesse's game against AZ Alkmaar two weeks later.

All de Mos was doing, though, was recognizing that football is a game defined by its weakest links. He was changing his brittle O-ring and hoping that Megrelishvili's replacement would perform at a slightly higher capacity than the hapless Israeli. He knew that leaving his bumbling defender on the field would have an enormous negative impact on his side's chances of winning either game. He probably did not want to humiliate one of his players. But then, he had no choice.

For all the money that is spent on superstars, there are limits to how great an impact they can have on any given game. In this respect the professional game is quite different from the amateur game. In a kick-about in the park the side with the best player or two will win almost all the time. Professional footballers share another similarity with those elite and skittish Prussian horses from our earlier chapter in that they have been culled from a large herd of eager youth players, promising teenagers and exceptional age-group talents. In a very Darwinian sense, the selection pressures are tremendous and the best players define the very limit of maximal fitness and skill. As they are selected from millions of candidates, they pile up right against this limit, which is determined by technology and science, as well as the physical bounds on sprinting speed, endurance and reaction times. This means that the spread of talent on a professional pitch is so much narrower than that in the park, and the ironic effect is to make outstanding players relatively less outstanding.

Moreover, even those players who can shoot hardest, pass most accurately (or, rather, create easy passes in tough situations), sprint quickest and run furthest must then come to terms with the fact that they will only have the ball at their feet for just 1 or 2 per cent of the time they are on the field of play.[1] This is another critical difference from the park game, where one or two excellent players can dominate possession. It distinguishes football from other sports such as basketball, baseball and American football, where the point guard, the pitcher and the quarterback have control of the ball for a significant portion of the contest.

No wonder our figures bear out the idea that it is the strength of a football team's weakest link that determines how much success a side will have, or that games are more often decided by errors, breakdowns in communication, or by finely tuned tactical systems falling apart. Football games are defined by mistakes; it is only natural that the worst player on the team is most likely to misplace a pass, or forget to mark his man, and lay a whole's week's preparation to waste.

It is the manager's job to minimize the potential impact of his worst player, both on the pitch on any given day and over the course of a season. Recognizing that football is a game disproportionately influenced by its weakest links is the first step: it should play a significant role in setting the manager's agenda.

To help every coach out just a little, we think there are five general plans available for solving a problem like Megrelishvili.

Understanding these sheds light on the significance of red cards, the importance of tactics, how and when to substitute and, possibly most importantly, the value of the superstar signing.

An orientation towards the weakest link might frustrate supporters. It means a manager knowing that, come the opening of the transfer window, he needs to spend more time and money seeking the perfect replacement for his own personal Megrelishvili than acquiring a crowd-pleasing marquee signing. However maddening that might be, remember that improving the weakest link is the most effective way to win more matches and climb the table.

Option One: Pretend He Doesn't Exist and Hide Him

Suppose we have a team with ten excellent players and one weak link where the substitutes' bench is packed with even feebler alternatives. Presumably, each starting XI is the best (least bad) XI. What is a manager to do?

In youth football there is one easy solution: you put your worst player in the position where he can do least damage and instruct your other, competent players, to ignore him. Footballers, being competitive animals, will probably do this instinctively; Steven Gerrard, when playing with Fernando Torres and Xabi Alonso at Liverpool, would always check to see where either of those two players were before passing to, say, Nabil El Zhar, the club's less-than-electric Moroccan winger.

Most managers faced with our dilemma would probably approve: didn't AC Milan, in Arrigo Sacchi's five versus ten training drill, effectively hide half a team of lesser players and never do worse than secure a goalless draw? True, they had possession rules in their favour and it was only in training, but they proved nonetheless that you don't need a full complement of players to get a result. Doesn't this show that sidelining a player is an option, one that enables a crafty manager to upgrade his weakest link to one of his ten excellent players?[2]

There is a flipside. By hiding him, the manager has transformed a player of some talent, however modest, into little more than a fan with a particularly good view. In actual fact, he might have taken his weakest link's output of, say, 40 per cent and turned it into a big fat zero. This might demolish the club's overall production in a multiplicative process. It must be better, surely, to have the galoot moving and participating in his own inimitable fashion than to have no player at all?

Fortunately, football has one situation that provides us with a decent test of whether a weak link should be sidelined or played – red cards. When someone is sent off, one player is now completely hidden in the dressing room, makes no contribution to the team's production, and eleven players magically become ten.

Red cards, like all important events in football matches, are rare. In Spain a team picks up a sending off about once in every five games; in Italy, that's once in every six matches and in Germany and England it's only once every twelve or thirteen games.

It would seem that the chances of that red card being received by a team's worst player are one in eleven: but that almost certainly is an underestimate, given that the worst player is also more likely to dive in late for a tackle, use his arm to flick away a header or be forced to pull a shirt to compensate for poor positioning.

Quick and easy calculations of the season-long performance

of red-carded players using data from Opta Sports show that, on average, players who receive red cards shoot less accurately, make fewer passes, a lower proportion of which are so-called 'key' passes, and commit more than twice as many fouls as players that don't receive red cards. We can, then, use red card data to give us an idea of what would happen if a team chose simply to sideline its worst player, to shove him off the pitch and tell him not to move. If it is a viable proposition, we should see that teams perform equally well, if not better, in matches in which one of their players received a red card.[3]

They do not. Looking at games from the big four European leagues over a number of years, we see that sendings off are damaging. Very damaging.

In Spain, England and Italy, receiving one red card reduces a team's point expectation for a match from about 1.5 to somewhere around 1, a reduction of a third. In the Bundesliga, over the five seasons from 2005 to 2010, a single red card cost a team almost half of its expected points, slicing 1.42 points per game with no red cards to 0.75 with one card. Red cards are very costly – playing ten-against-eleven football is a recipe for defeat.[4]

There are different sorts of red cards, of course. Luis Suárez's dismissal in Uruguay's 2010 World Cup quarter-final against Ghana for handling Dominic Adiyiah's header on the line was a proxy for poor defensive play. It was also a blatant example of cheating. But, more importantly to Uruguay, it also traded up the certainty of defeat had Adiyiah scored to a 75 per cent chance of losing if Ghana converted the subsequent penalty. It was, in that sense, a calculated gamble, and one that paid off.

Then there are red cards that are sustained in the heat of the match – think Zinedine Zidane on Marco Materazzi in the 2006 World Cup final, or Wayne Rooney on Ricardo Carvalho in the quarter-finals of that tournament. These are, it would seem reasonable to suggest, more common in games when your team is condemned to defeat, when things are going badly and frustration has set in, or when a player is lashing out in response to provocation from the other side. In statistical terms, our simple test above might be biased towards showing a big negative impact from red cards. This means we need to apply a more sophisticated analysis if we are to confirm the negative effects of red cards and the harm of simply hiding the weakest link.

By running a regression on data from all four leagues for the five most recent seasons, while accounting for match-specific differences – home advantage, shots, goals and fouls – we can show the connection between the number of red cards and the likelihood of a team losing or winning a match.[5] Here, too, it is clear that red cards increase a team's chances of defeat.

Going from no red cards to one increases the probability of earning no points from 24 per cent to 38 per cent. If your team gets a second red card, losing becomes the most likely outcome, even more probable than gaining a point. The chances of taking three points decreases from 36 per cent to 22 per cent when a team has a man sent off, and the odds against winning to more than 7–1 when a team has two players dismissed.

The most fitting comparison is with being at home: playing on familiar territory, as opposed to at an away ground, increases a team's chances of winning from 27 per cent to 42 per cent, while decreasing the chances of losing from 32 per cent to 19 per cent. A single red card costs a team 0.42 expected points, while changing where a match takes place from home pitch to away costs a team 0.43 expected points. Having a man sent off is roughly the same as giving your opponent home advantage.[6]

Removing your weakest link entirely, then, hiding him in the safety of the dressing room, is simply not a risk worth taking. But would it work to play him in a position where he can do the least damage? Traditionally, there has only been one place where poor players are stationed: right and left back, the Elbas of the pitch. Jonathan Wilson describes Gianluca Vialli's theory that 'the right back is always the worst player on the team', because the good defenders are moved into the centre, the good ball-players are moved into midfield and the left-footers are so rare that they have to be nurtured.[7] Simon Kuper, on the other hand, believes that 'nobody cares about left-backs'. His example is Roberto Carlos, one of Real Madrid's *galácticos*, who 'passed largely unnoticed until the age of twenty-four'.[8]

Perhaps, years ago, a manager would have been able to get away with hiding his worst player at full back, but with the rise of video analysis, extensive scouting and a more intensive pace of the game, it seems unlikely that a team could conceal a weak link for long. Look at Arsenal, who saw Gaël Clichy, their left back, endure a horrible period of form in the second half of the 2009/10 season. When he faced Manchester United in a home fixture at the end of January, a match the visitors won 3–1, he was tortured by the pace and power of Nani, the visitors' Portuguese winger. This was no accident: Michael Cox of the tactics website Zonal Marking noticed that United's goalkeeper, Edwin van der Sar, had placed the vast majority of his goal kicks towards the area patrolled by Clichy (Figure 46).[9]

Arsenal couldn't hide Clichy and had to counteract United's targeted bombardment. William Gallas, the central defender, edged left to cover his full back. Cesc Fábregas and Samir Nasri

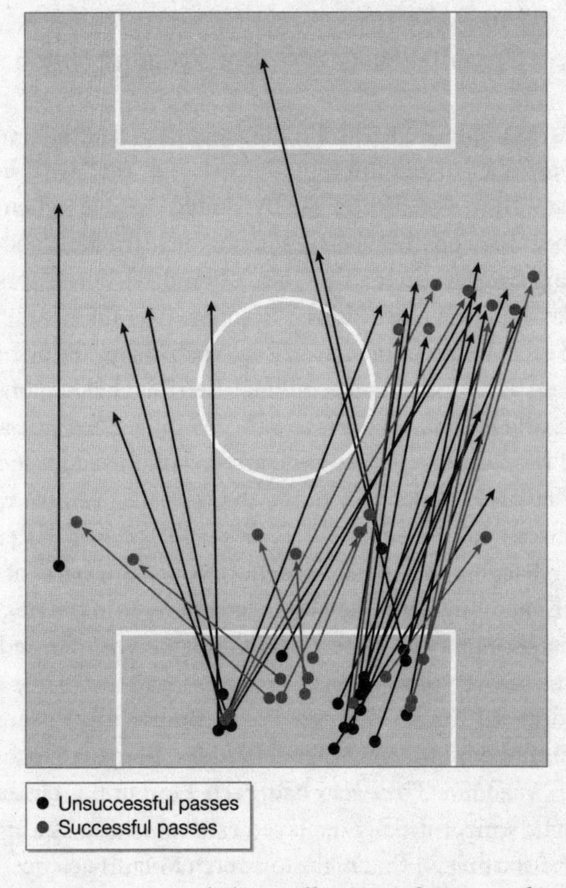

Figure 46 Passes made by goalkeeper Edwin van der Sar for Manchester United against Arsenal in January 2010

dropped further back to add support.[10] This is what we would call the 'finger-in-the-dam solution': you improvise and reinforce the weak spot with whatever materials you have on hand. This is the second option open to a manager: if you have a weak link, get your other players to help him out.

Option Two: Face Reality, Reinforce Him

In that example from the Premier League, Arsenal's off-the-cuff attempts to reinforce their weak link met with limited success – they were beaten 3–1 by United. Indeed, when weak links become apparent during a game the team must concoct a covering strategy on the fly. Quite often these don't work.

Take 'La Quinta del Buitre', five players at the heart of one of the most lionized line-ups in Real Madrid's glittering history. In the 1989 European Cup semi-final, Leo Beenhakker, Madrid's Dutch coach, took his team to San Siro to face Arrigo Sacchi's AC Milan. The first leg, in the Spanish capital, had finished in a 1–1 draw two weeks previously, thanks to an extraordinarily fortuitous goal from the Dutch striker Marco van Basten. His header had hit the crossbar, deflected off the back of goalkeeper Paco Buyo and then bounced slowly into the net.

This, though, was quite a Madrid team. They fancied their chances of overcoming Milan even on home turf. They had at their disposal Emilio Butragueño, El Buitre, the vulture, and his four cohorts: Míchel, Miguel Pardeza, Manolo Sanchís and Martín Vázquez. They also had Paco Llorente, a blisteringly fast right winger usually deployed as a substitute but brought into the starting XI that night to stretch Milan's defence.

The plan backfired spectacularly: Butragueño ended up being forced wide to shore up the right flank, disabling his partnership with Hugo Sánchez, while Bernd Schuster, Real's midfield player, could not make any impact at all on Frank Rijkaard and Carlo Ancelotti, who ran the game for the hosts.[11] Milan found space abundant and met weakened resistance. La Quinta del Buitre were stripped to the bone, losing 5–0.[12]

When the weak link is reinforced not through improvisation but through well-planned strategy, results tend to be rather more impressive. One of the most famous formations in football history – *catenaccio* – was based on this principle.

As David Goldblatt explains in his authoritative history, *The Ball Is Round*, *catenaccio* as a system of play was first developed by the Austrian-born coach Karl Rappan at Servette in the 1930s.[13] His innovation was to withdraw a player from his forward line and play him behind his three centre backs. He had no direct opponent to mark; instead, he would protect space.

The ploy worked spectacularly: Servette and then Grasshoppers brought Rappan seven Swiss league titles during the 1930s.

As is true of many innovations – the motor car, proving that the Earth circles the sun, television – there were independent development efforts in other locations, particularly Italy, for finding ways to manage weak links with different tactical formations. Gipo Viani, the manager of Salernitana, went from Serie B to the heights of Italian football as manager of Roma, AC Milan and the Italian national team after a flash of inspiration by the Salerno docks which convinced him to deploy a sweeper centre back:

> Oblivious to the shrieking of the gulls and the haggling of the dockside mongers, he strides on, asking himself again and again how he can get the best out of his side, ponders how he can strengthen a defence that, for all his best efforts, remains damagingly porous. As he paces the harbour, churning the problem over and over in his head, a boat catches his eye. The fishermen haul in one net, swollen with fish, and then behind it, another: the reserve net. This is his eureka moment. Some fish inevitably slip the first net, but they are caught by the

> second; he realizes that what his side needs is a reserve defender operating behind the main defence to catch those forwards who slip through.[14]

It would be Nereo Rocco, at AC Milan, and Helenio Herrera, at their city rivals Internazionale, who would develop this strategy most comprehensively. Between them, they forged a system that would define Italian football for two generations at the very least. Under their guidance, *catenaccio* became identified with a brutal, cynical, defensive, inelegant, cautious style of play. As Rocco famously told his players: 'Kick everything that moves; if it is the ball, even better.'

That should not, however, be allowed to cloud what *catenaccio*, in its original form, was meant to be: a way of solving football's most significant structural problem – protecting your team's weak links.

Option Three: Substitute Him

Not all managers can be a Viani, a Rocco, a Rappan, or a Herrera. Not all managers can traipse around fishing boats at night coming up with elaborate tactical schemes. And not all managers trust their players to cover for one another, or to make up for the shortcomings in their selected systems. Most managers prefer to be in control. And that means following Aad de Mos's lead – though usually in slightly less brutal fashion than he employed with Megrelishvili – and identifying and removing your weak link.

Sounds pretty easy. Watch your team play, work out which player is doing least well and send on a replacement in his stead. There is far more to the art of substitution than that, though, as research over the last decade has proved.

Academics at the University of Oviedo and the Technical University Lisbon have found the vast majority of players who exit the pitch for a substitute are midfielders, and 40 per cent of all substitutions are midfielder for midfielder.[15] Most forwards are replaced by other forwards, but again almost 40 per cent of them are replaced by midfielders. Defenders are replaced the least, and defenders and forwards are very rarely swapped for each other.

According to Bret Myers, once a player of note for the United Soccer Leagues' Richmond Kickers and now a Professor of Management and Operations at Villanova University, only a small fraction of substitutions happen in the first half – and very few happen in the first six minutes.[16] His sample of games from the Premier League, La Liga and Serie A shows that most first substitutes come on at half-time and between the fifty-sixth and sixty-fifth minutes of the match; most second subs are used between the sixty-sixth and eightieth minutes, and the third substitutes happen in the last ten minutes (plus injury time) of the game (Figure 47).

Is this the best way of using substitutes? Is there a way not only of replacing your weakest link, but replacing him to maximum effect?

Myers has an answer. Taking the same sample of matches shown in Figure 47, he used data-mining techniques to test whether a particular minute, from forty-five to ninety, was the critical moment for distinguishing good substitutions from those that were likely to be too late to affect the final outcome.

The statistical software could test whether managers whose

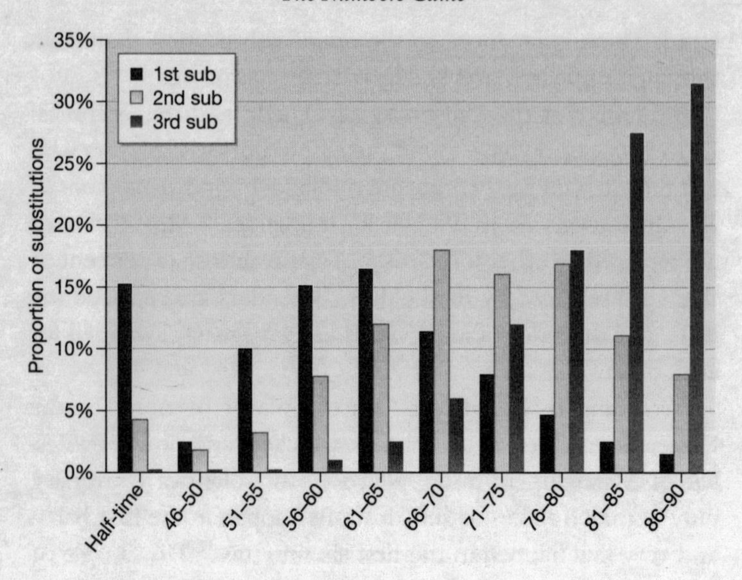

Figure 47 Timing of substitutions

teams were drawing and who substituted a player at half-time were more likely to have avoided defeat than those who first substituted at a certain minute after the break – the forty-seventh, forty-eighth and so on. It could also test whether it mattered if the club was behind, level, or ahead when the manager made the substitution.

Myers turned this into a pseudo-experiment by taking the substitution rules that were most predictive of success in the sample and then testing them against a much bigger selection of match results that included the Bundesliga, the World Cup and Major League Soccer. What he discovered was nothing less than a recipe for substituting success, a handbook on the art of replacements.

According to his findings, if a manager's team is losing, for maximum effect he should make his first substitute before the fifty-eighth minute, his second before the seventy-third, and his third before the seventy-ninth. If he is not losing, it does not matter when he makes his substitutions.[17]

If this strikes you as being slightly aggressive, remember that we are looking for ways to manage our weakest link. Each player has an expected level of performance and it is fair to say that those players who have the better expected performance levels start the game. That will reduce over the course of the game, and at some point a substitute's expected level of performance will start to exceed that of his tired and ineffective teammate. This is when to make your substitution.

This is not a rule that many managers seem to follow, as Figure 48 reveals. The stair-stepping line down the centre of the horizontal bars divides those matches in which the manager followed the rule from those in which he didn't; moreover, the bars from top to bottom are ranked by frequency of

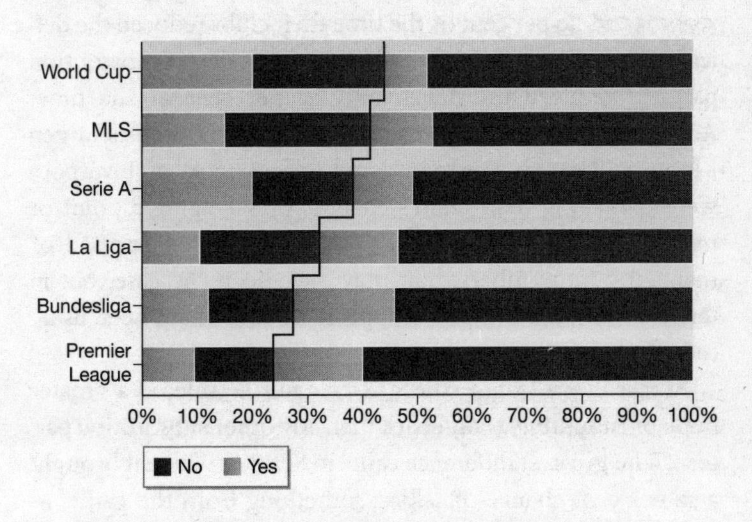

Figure 48 Effect of substitutions on score deficit

following the rule. More managers adhered to what we might call the <58<73<79 principle in the World Cup and MLS than in any other league, with managers in the Premier League and the Bundesliga revealing themselves to be the most conservative substituters. Some 44 per cent of national teams in a World Cup substituted aggressively when they were trailing, while less than a quarter of losing Premier League teams did.

An additional piece of information is contained in the figure. The shading within the bars distinguishes those matches in which teams managed to diminish or erase the deficit after substituting ('Yes') from those in which the score stayed the same or moved in even greater disfavour ('No'). As the relative proportions of dark to light shading within the bars reveals, conservative substitution by managers in the Bundesliga or Premier League may have cost their teams points.

Across the Premier League, when managers followed <58<73<79, 40 per cent of the time their clubs reduced the deficit and often drew the game, whereas other, slower substitution patterns reduced the deficit only 22 per cent of the time. According to this theory, reluctant substituters such as Jürgen Klopp of Borussia Dortmund and Rafa Benítez of Liverpool were damaging their team's chances of salvaging a point or more from a losing position. Benítez was often accused of managing by numbers; that may have been the case, but in terms of his substitutions, it appears he could have been using the wrong numbers.

Across every league, the <58<73<79 rule offered a greater hope of mounting a comeback than any other substitution pattern. The greatest difference came in Serie A, where it brought a 52 per cent chance of taking something from the game, as opposed to 18 per cent if the rule was not followed.

Those managers who never followed it paid the price. Take Tenerife's José Luis Oltra, who never once used the <58<73<79 rule to replace his starters. They managed just two comebacks in games where they were trailing, were duly relegated and their manager sacked.

The reason some managers fail to follow the rule is obvious: they simply misjudge the crossing point where players' objective performance declines sufficiently to warrant replacement. But *why* are managers bad at substituting at the right time? Psychologists wouldn't be surprised. Delay is a normal human decision-making bias: managers are committed to their initial assessment of the gap in performance (they 'trust' the starter). This commitment – it's called anchoring – means that they have a hard time assessing the more uncertain crossing point, so they wait until the evidence of flagging performance is undeniable. By that time, it's often too late. The manager should therefore substitute *before* his eyes and brain tell him to.

To compound matters, players do not usually want to be removed unless they are injured. They are experts at making a manager believe they have plenty more to give, which makes judging their performance levels even more tricky.

One study, by Lille's sports scientist Chris Carling and colleagues, used a computer system to track things like the distance players covered, the intensity and frequency of sprints and the time they need to recover from high-intensity efforts.[18] Carling and his team found no differences in the first- and second-half performance levels of players who were substituted.

This seems to suggest that there is no drop-off in physical performance, that there is no reason to substitute. Not so: another study co-authored by Carling that looked at players' work-rates after a teammate had been sent off shows that footballers are experts at pacing themselves and operating at less than full capacity.[19]

As a result, fatigue may not show up in their average work-rate; it may not be visible from the dugout. Instead it shows up when they try to go from the 90 per cent capacity they are operating at to the 95 per cent they need to stretch for a tackle or to leap for a header. Early in the game, they can do this easily. Later, when they are pacing themselves, reaching the required capacity is no longer possible.

The critical piece of evidence, then, might be not how the substituted players were performing on the pitch, but how their replacements performed. Carling and his co-authors found that, while forwards who came on as substitutes also seemed to be pacing themselves, midfield substitutes – remember, by far the most frequent replacements – 'covered a greater overall distance and distance at high intensities and had a lower recovery time between high-intensity efforts compared to other midfield teammates who remained on the pitch'.[20]

In other words, players who come on perform at a higher level than players who come off. If a manager waits for clear signs of fatigue, he might be substituting too late: he might make better decisions by following a set rule such as <58<73<79.

During his ill-fated reign at Portsmouth in the 2009/10 season, Avram Grant could have followed that rule in twenty-one games. He chose to do so only four times. On two of those occasions, his team made a dent in their deficit. In the seventeen other games, where he did not follow that rule, the deficit stayed the same or increased fourteen times. If he'd followed the rule, Pompey's results might have been sufficient to avoid relegation.

Option Four: Try to Improve Him

A really good manager will take his weak link under his wing, give him the benefit of all his wisdom and make him a better player. The way we think about it, during the week, managers are essentially employed to do two things: develop tactics to try and conceal the extent of their weakest links, and coach weaknesses out of players. In broad terms, there are two categories of weakness – effort and skill. The first requires the manager to motivate, and the second to teach.

A: Get Him to Work Harder

In addition to inspiring speeches, kicks to the seat of the pants and punishing drills, the smart manager, often unwittingly, will use the Köhler effect to increase the effort of his weak links.

This phenomenon is named for Wolfgang Köhler, head of the Psychological Institute of Berlin University in the 1920s. Köhler is an inspirational figure: he had built quite a team, a *galácticos* of modern psychology, but saw them disbanded when the Nazis rose to power in 1933. Many of those he worked with left Germany for America, while his Jewish colleagues were stripped of their positions.

Köhler, though, did not suffer in silence. In April 1933 he wrote the last German newspaper article to criticize the Nazis until Hitler's death twelve years later, and subsequently he greeted the mandatory introduction of the Nazi salute at the start of lectures by telling his students – some of whom, no doubt, were party loyalists – that he 'did not share the ideology which it usually signifies or used to signify'.[21]

Given his work, it should be no surprise that Köhler possessed the willpower to stand tall out of loyalty to his scattered and suffering former colleagues.

Through a very simple series of tests performed on members of the Berlin rowing club, Köhler had demonstrated that teamwork could produce significant gains in motivation. First, he tested how long each standing rower could, while holding and curling a bar connected to a weight of forty-one kilograms, keep the weight from touching the floor.

Then he doubled the weight, paired the rowers and tested how long they could curl the heavier bar together. This is a weak-link task because the weight was too great for any single person to hold up: the eighty-two kilograms would hit the floor when the weaker partner's biceps gave out. Köhler found that weaker rowers would endure significantly longer when they were paired than when they were solo. In doing so he had isolated one of the key characteristics of psychology: the gain in enthusiasm and effort and perseverance that comes from being on a team.

It was not until the 1990s that psychologists began to investigate the reasons behind Köhler's findings. They found two causes for the effect: a social comparison process, where individuals perform better when working with a more capable partner, and an 'indispensability' condition in which individuals do not want to hold back the group, and feel that their contribution is crucial to collective performance. Or, to put it more bluntly, the Köhler effect occurs because weak links work harder to keep up, whether in an attempt to match their more talented colleagues or because they think their role is just as essential. These two factors are equally important in helping improve a weak link.[22]

There is plenty of evidence that this applies to the world of sport. Take Jason Lezak, the final swimmer in the American 4×100 freestyle relay at the 2008 Olympics. Lezak was not the strongest swimmer in the team – despite being in probably the most crucial spot – and found himself up against Alain Bernard, anchor of the French team and the 100m freestyle world record holder. Worse still, when Lezak's leg started, he was a full body length behind.

No matter. Lezak swam the fastest 100m relay split in world history, in 46.06 seconds, securing the gold for the USA. His split was faster than Bernard's by 0.67 seconds, and he out-touched the Frenchman by 0.08 seconds, an eye blink.[23] The reason Lezak, who had won just two individual medals in major competitions, gave for his superhuman performance? 'I'm part of a team, and today was no different. I got with the guys and said, "We're not a 4-by-100 team. We're all one." '[24]

Lezak's not the only swimmer to display the Köhler effect. A recent study of all the relay teams from the Beijing Olympics provides dramatic confirmation of its existence.[25] Those athletes who swam the second and third legs beat their individual times by 0.4 per cent on average, and those who swam the anchor legs, the most indispensable, beat their solo times by 0.8 per cent, both significant margins in a sport decided in blinks of an eye. Köhler's weightlifting rowers did discover a very real phenomenon that can affect all teams, in business and in sports, including, of course, football.[26]

That's not to say that harnessing the Köhler effect is easy. It would mean a manager convincing a player earning millions, with a sycophantic agent and surrounded by an admiring entourage, that he is the worst player in the squad. That would be an interesting conversation, though it is not an impossible one. An adroit manager would, perhaps, be able to foist blame on a recent injury, a tough run of opposition, or his own failings as a boss to ease the player's dissatisfaction. Then the manager has to make the unfortunate player believe he can improve, and show him a path and a training programme that seem promising. In addition the manager has to promote the philosophy within the club that football is a weakest-link game, and that therefore everyone's contributions are essential.

This may be easier in some teams, such as military special forces units, where differences among teammates in compensation and recognition aren't dramatic, than in modern football. While the game says that everyone is essential, the salaries say that some are 'more essential' than others.

Crucially, the manager has to be blessed with a team-minded strongest link, the sort who is first to the training ground in the morning and last to leave at night, promoting an ethos of maximum effort at the same time as making it easier to think that his success is down to hard work, not sheer talent. This is better in terms of motivating the weak link, who can match the strong link's effort, if not his virtuosity.

There are countless examples of strong links who do not fit this mould. Allen Iverson, the basketball superstar, held a famous press conference mocking the idea of practice and calling it 'silly'.[27] Paul McGrath, while at Aston Villa, or Ledley King at Tottenham were not nearly so cocksure, but the need to manage injury prevented them joining in training sessions during the week. There are countless Brazilian strikers – Adriano, Edmundo – who regularly chose not to train, relying instead on their inherent ability. The effect on players without their talent would have been profound: work all you like, you'll never be in my class. Not good for the team.

One superstar who does fit the mould is, no surprise, Lionel Messi. He's never viewed training as silly or unimportant. He wouldn't be caught dead repeating Iverson's scornful words at a press conference: 'We're sittin' here, and I'm supposed to be the franchise player, and *we're talkin' about practice.*' Messi's teammate Gerard Piqué observes, 'He could say, "OK, I'm the best, but in training I don't care, I can be lazy," but he's working at the same level in training as well. It's unbelievable.'[28] Surely, Piqué, very far from being a weak link, nonetheless goes a little harder in practice than he otherwise would.

B: Teach Him New Skills

Football is not a sport where effort matters more than skill; instead, technique, physical abilities and mental aptitude are at least as important. Many times the manager teaches his players, especially the weak links, directly. This could take the form of collective training: Xavi Hernández has detailed how Barcelona's practice sessions revolve around short passing exercises, so that even the worst technical player grows used to passing and moving, passing and moving. Some managers will even give certain players one-on-one tuition to improve a certain aspect of their game: Rafael Benítez, while at Liverpool, spent days teaching Ryan Babel how to vary his wing play more effectively.

As with motivation and increasing the weak link's effort, the manager need not do everything himself. Instead, he can structure the squad and create a club culture that fosters skill development. For clues on how to make this happen, we must venture for a time far from the pitch again, not to swimmers in a pool, but to seamstresses in a garment factory producing women's clothes.[29]

For years, workers at the Koret plant in northern California had sewn to an individual piece rate: they were paid five cents per belt loop, and the more belt loops they sewed, the more they would be paid. This was the way it had always been done, but it led to some inefficiency: the factory floor was full of carts of works-in-progress, as partially finished garments were moved from station to station, awaiting the next piece of fabric.

In 1995 the Koret plant switched some of its plant to 'module' production – establishing teams responsible for sewing entire garments. They would receive a single sum for each finished item that would be shared among all their team members. Koret's management thought productivity would fall (far fewer skirts would be made), but that this would be offset by reduced waste and higher quality, as more flaws were noticed more quickly.[30]

All the staff had previously worked under the individual system, and only some of them switched over to module sewing. This was the perfect laboratory – with before and after numbers, and a control group of sewers still under the piece rate – to test the effect of team membership on performance.

The results, found by the economists Barton Hamilton, Jack Nickerson and Hideo Owan of Washington University in St Louis, were just as impressive as they were unexpected. There was, on average, an 18 per cent *increase* in production, most of which was due to the team effect. Three teams exceeded the productivity of their best worker. The teams with high-ability workers were more productive, but so were teams with a great spread in ability.[31]

This last part is critical. Lower-skilled workers improved because of the Köhler effect, and because the better tailors shared their knowledge. In another factory, 90 per cent of module workers said that the 'informal training' they received as part of a team improved their work.[32]

Footballers are no different. Inspired partially by the Koret study, the Swiss economists Egon Franck and Stephan Nüesch examined the performance of Bundesliga clubs from 2001/02 to 2006/07.[33] They did not look exclusively at shots and goals, but instead used Opta Sports match data to create a performance index for each position, allowing them to calculate the average talent level for each match and club on the pitch, along with the spread in talent across the eleven players.

Their analysis confirmed both the O-ring theory and the presence of learning. In the short term, when Hannover 96 face Hamburg the match is decided by luck (of course), then by who's at home, then by which team has the higher average talent level and then by which club has the *narrower* spread of talent. It is better to have a team of all 70 per cent players than it is to have a team where two players are 100 per cent, the majority are 70s and then there is one bumbling 50 per cent and one dreadful 30 per cent. Strong links don't win matches. Weak links lose them.

Over the course of the season, though, the converse is true: where you finish in the table is determined by average talent level (Bayern Munich will end up higher than Kaiserslautern) and then by a *broader* spread of talent in the squad as a whole, rather than just the starting XI. For the season, in our example above, Hannover 96 would be better off with a squad including two superstars and two weak links, *as long as* the better players transform and lift the less able ones. As Franck and Nüesch state: 'A professional soccer player invests up to eight hours a day in soccer-related preparatory activities. Here talent heterogeneity should increase team performance, as it enables the less able players to learn from their more talented teammates. Furthermore, talent disparity also affects the social norm of productivity and the resulting peer pressure during training activities.'[34] In other words, as with the slow seamstresses in the Koret factory, the weak links on a football team can be inspired to work harder and taught to play smarter by the strong links.

The manager, then, should try to establish a club culture in which the weak links are willing to ask for help and will listen to advice. When he buys a superstar, the manager should realize that he is not just buying the goals and step-overs and back-heels, he's also buying a set of habits and attitudes, a willingness to help and a commitment to his teammates. These qualities may be as important as what the star does on the pitch, because of the effects on the weak-link players.

Option Five: Sell Him

Some weak links cannot be hidden or improved. Some players simply will not get better, no matter how much you try to help them. They will not learn from their peers, or be able to keep up with their teammates. Reinforcing them may weaken other areas of your side, and there are only so many times a player can be substituted. That leaves just one solution.

Every player leaves a club sooner or later, whether it's for money, for ambition, for age, for dwindling ability, or simply for a change of scenery. The decision usually rests in the hands of one man: the manager. This brings with it its own risks. The manager might say he has done all he can with his worst player. Assuming the player is saleable, the manager might say he needs the funds to spend on a new signing. But whether he is correct or not rather depends on his own ability. For a club to know selling a player is the right decision, it must be sure that it has the right manager.

Eight months after his ignominious early replacement,

Haim Megrelishvili was placed on the transfer list by Vitesse. There was one small victory for our hapless Israeli, though. He was not sold by Aad de Mos. Vitesse's weak link had at least outlasted his manager, whose contract had been terminated a mere six weeks after the sixth-minute substitution. The club had decided that de Mos was even more of a problem as a manager than the weak link was as a defender.

10.
Stuffed Teddy Bears

> I am the very model of a modern major general,
> I've information vegetable, animal and mineral,
> I know the kings of England, and I quote the fights historical
> From Marathon to Waterloo, in order categorical,
> I'm very well acquainted, too, with matters mathematical,
> I understand equations, both the simple and quadratical.
>
> *Gilbert and Sullivan*

José Mourinho might lay claim to being the greatest manager in the world. The Portuguese is just one of three men in history to win the European Cup with two different clubs, and the only man to do so in the competition's Champions League era. He is one of only four coaches to win league titles in four different countries, and he is the only man on both lists. Whether it was at FC Porto in his homeland, at Chelsea, at Internazionale, at Real Madrid or back at Stamford Bridge, like him or loathe him, Mourinho has a golden touch.

But then so does Sir Alex Ferguson. He picked up twelve Premier League titles in his time at Manchester United, as well as two Champions League trophies, five FA Cups, four League Cups, the European Cup Winners' Cup and the Fifa World Club Cup. He remained in place at England's largest club for more than a quarter of a century. Surely that durability must count for something, putting him ahead of the bright, but brief, flame that is Mourinho?

In that case, maybe Jimmy Davies should be considered for the title. That's right, not Carlos Bianchi, who forged one of the finest teams of the modern age at Boca Juniors, or Pep Guardiola, inspiration behind Barcelona's domination in recent years, or Marcello Lippi, World Cup winner with Italy and Champions League winner with Juventus, or Vicente del Bosque, who managed the same trick with Spain and Real Madrid, or Fabio Capello, or Marcelo Bielsa or Arsène Wenger or any of the other usual contenders. When it comes to longevity, none of them are a patch on Jimmy Davies, the manager of Waterloo Dock AFC, a non-league team on Merseyside.

Davies has all the hallmarks of a great manager. He's a straight-talker, unafraid to tell his players when they've underperformed. He's a micro-manager in the truest sense: he makes sure the corner flags are correctly planted, his squad's shirts hung neatly on their pegs, and he fills in the team-sheet himself. The method obviously works: he's won twenty-eight cups in his time at Waterloo Dock, as well as twenty-one league titles, including five straight between 2007 and 2011. He is, by his own admission, a nurturer of talent. 'Some of the finest footballers on Merseyside have passed through our ranks. Our list of honours bears witness to this ability. The future players of today and tomorrow have a tremendous task ahead of them if they are to emulate the feats of their predecessors. For our part we never fail to remind them of our history and the expectations we place upon them.'[1]

He is also the longest-serving manager in the illustrious history of the Football Association. He led Waterloo Dock for fifty years before his retirement in 2013; he has a quarter of a century on even Ferguson, the standard-bearer for gnarled and wizened old masters.

Before we dismiss the minor-league manager's case out of hand in favour of Ferguson, consider the similarities: the burden of a glorious history, the demand for results, man management across a wide generation gap. Of course Davies did not have to deal with major-league pressure: the relentless media attention, the superstar egos and the constant challenge to his primacy that Ferguson endured at Old Trafford. But then Ferguson did not have to put up with the travails of minor management: a crippling lack of budget, empty stands, an ego-shattering dearth of attention and the loss of his best players to overtime shifts in their full-time jobs.

Given these vastly different environments, it can seem impossible to compare managers. What might Davies have done had he been given the chance at Manchester United? Would Ferguson have coped with the rigours of coaching a team, never mind hand-washing the kits, while buried deep in England's non-league pyramid?

This issue raises its head with enormous frequency across the world. The manager's position is totemic, and with good reason. In his hands rest most of those decisions that can influence that part of a team's fortunes not determined by luck. He must decide how to handle the weak links, he must strike a balance between the light of attack and the dark of defence, he must find a way to cope with football's multitudinous and glorious inefficiencies. He must secure as many of those beautiful, rare goals as possible. He is cast as the modern major general, the centre of power in football's galaxy.

And yet there is no universal measure to assess how good he is at his job. Is it tenure? Is it cups or titles? Is it the undying support of fans? One strong answer is that there is no universal measure: who the best manager is does not matter, because the manager himself is irrelevant. This modern opinion on leadership declares that the manager has all the significance of Gilbert and Sullivan's Major General Stanley and all the impact of a stuffed teddy bear. That is to say: none. Before arguing over the qualities that make for an excellent manager, we have to establish whether leaders – Mourinho, Ferguson, or Davies – matter at all in the first place.[2]

The Anti-Cult of the Manager

It was in Chelsea that the theory of the Special One was first born, long before Mourinho turned up at Stamford Bridge. In 1840, a few years after he had left his native Scotland for west London, Thomas Carlyle wrote that 'the modellers, patterns, and in a wide sense creators of whatsoever the general mass of men contrived to do or to attain' were the Great Men of history. 'All things we see standing accomplished in the world are properly the outer material result, the practical realization and embodiment, of Thoughts that the Great Men sent into the world.'[3]

To Carlyle, heroes make the world and drive history: from King Arthur, who pulled a sword from a stone to found a kingdom, to Martin Luther, who brought the Reformation into reality. The rest of the general mass of men, he told us, would always retain 'admiration, loyalty, adoration' to these colossi who stride over the age of man.[4]

His theory has since been discredited, but he identified a thread in Western culture and history. For all the enlightened nations that profess a loyalty to liberty, democracy, economy and all the rest, there has long been a readiness to look for a chosen one; as Carlyle pointed out, even the French, those great anti-venerators, those relentless beheaders of Great Men, worshipped Voltaire, even 'plucking a hair or two from his fur to keep as a sacred relic' upon his visit to Paris.[5] Humans, he knew, do quite like a hero, a sage, someone who knows best. We are, to some extent, always searching for our own Special One.

This is particularly true of football managers, subjects of some of the most intense hero worship in the modern world. As Barney Ronay observes in his book *The Manager: The Absurd Ascent of the Most Important Man in Football*, the manager's 'shadow looms, fully evolved, now erect and miraculously walking on his hind legs: priest, messiah, hard-nut, patriarch and visible emblem of over a hundred years of confused and piecemeal progress'.[6] It's a swift give-and-go that connects Carlyle's heroes to those in the dugout – Napoleon to Roux; Luther to Rehhagel; Burns to Ferguson; Cromwell to Chapman; Shakespeare to Shankly; Dante to Trapattoni; and Jesus to – need we even say it? – Clough.

Irrespective of nationality, it was in England that the Great Men of football were first forged. A simple glimpse at the vocabulary of the world game offers proof enough of this: across football's heartlands in South America, Italy and Spain, managers are still referred to as 'Mister'; England sent out missionaries to Eastern Europe (Jimmy Hogan) and Scandinavia (George Raynor). Ronay describes the manager as England's 'gift to the wider world beyond its boundaries'.

The idolization was no less intense in the motherland. Arthur Hopcraft, in his newspaper columns and his 1968 book *The Football Man*, captured the rise of the manager perfectly. 'To watch Sir Matt Busby move about Manchester is to observe a public veneration,' Hopcraft wrote. 'He is not merely popular; not merely respected for his flair as a manager. The affection becomes rapidly more deferential as they get nearer the man.'[7]

This is true across the world. The heroic manager wins cups, delivers promotions, attracts glory and fame. Only poor fortune or slapdash players can derail him from arriving at his destiny.

There are always doubters, though, those who see false prophets. Carlyle identified this tendency, lamenting how the mean and mediocre reduce the Great to mere circumstance: 'Show our critics a great man, a Luther for example, they begin to what they call "account" for him . . . and bring him out to be a little kind of man! He was the "creature of the Time", they say; the Time called him forth, the Time did everything, he nothing – but what we the little critic could have done too!'[8]

Managers are no exception. Anti-cultists insist that they are worthless, that their impact is minimal, that their decisions are superfluous to the overall result, that it is the players who drive history, and that the managers are simply there to ensure they stay fit and know their teammates' names. There are those – maybe not that numerous, but they are there – who swear the emperor has no clothes.

This anti-manager zealotry springs from two sources. The first is a pure Jacobin, anti-authoritarian compulsion. It compels us to want to pull down Great Men, to call for their heads and declare them incompetent. It happens in all walks of life – politics, public service, business – to declare the leaders inept and hapless. In football, it tends to be particularly public and unimaginably vicious, as Graham Taylor, labelled 'Turnip Head' by the *Sun* newspaper for his failure to help England perform well in Euro 1992 or even reach the 1994 World Cup and, eventually, hounded from his job, can attest.

The second is more recent still, and rests upon a change in broader culture wrought by the media, computers, the internet and our own sad, vacant lives – the rise of the celebrity and the video game. As Ronay writes: 'In the early 1990s, football entered a new era. A media-led, lad-culture-infused revival was in train. Football was cautiously on its way to becoming a mainstream pursuit, a lifestyle choice in an era of aggressively marketed leisure. The manager was part of the wider scene now. There was no need for him to seek fame. It came looking for him.'[9] The manager was no longer the hero of Busby's times, but celebrity. Heroes exist in the firmament and create awe; celebs are just like us. They are written about, gossiped about and commented on ad nauseam.

This has been exacerbated by the rise of the football management simulation. Games like the *Football Manager* series have reduced what was once seen as a special skill possessed only by a few to little more than an algorithm for success. Now anyone can simulate how they would do if they were manager of their favourite club. They can see, on a screen in front of them, quite how their genius would take York City to the Champions League final. They know they could have done what their manager does, if only they had been given a chance.

The algorithm at the heart of these simulations has grown more complex with each version of the game. More statistics and numbers have been incorporated into the code so that you have more finely grained decisions to make about more aspects of the club. The game seems to get more and more real. *Football Manager* makes it seem that all decisions – the salaries and bonuses, the team meetings, the formations, the training sessions – are essential and impactful, but that making these decisions is as easy as clicking a mouse. Any 'little critic could have done too'.[10]

Suddenly the manager is not nearly as venerated as he once was. He is a figurehead still, but one to be mocked and lampooned by those who think they know better. The age of the Great Men appears to be over. Everybody knows what it takes to make a manager, and where the current incumbent is going wrong. Or, at least, they think they know.

Football Accountants (Reprise)

Perhaps the most compelling evidence of the *un*importance of the manager comes from work by sports economists on the strong correlation between wages and wins in football. What matters more than who's on a club's team-sheet, their thinking goes, is what sort of figures are on your spreadsheet.

Simon Kuper and Stefan Szymanski are probably the most prominent. In *Why England Lose* they came bearing evidence as to the futility of the manager. They show that, for the decade beginning in 1998 in the Premier League and Championship, a club's total spend on wages explained 89 per cent of the variation in its average place in the final league table.[11] For the decade, Chelsea, Manchester United, Liverpool and Arsenal were the top four (in order) in total wages paid, and their average table finishes were third, first, fourth and second, respectively. Crewe Alexandra, Brighton and Rotherham paid the smallest salaries, and their average finishes were tenth worst, worst and third worst of all clubs who played in the two divisions during that period.

As a standalone figure, this is pretty damning: 89 per cent of where a club finishes is determined by its accountant. All the effort the manager puts into training, devising tactics, screaming in players' faces, employing puerile mind games in the media: all this is over that puny 11 per cent that money can't buy (and, if we factor in the role of fortune, this can probably be reduced to 5.5 per cent). The key figure in football is not the manager, striding around the technical area like he owns the place, but the paymaster, the man who actually does.

The idea of manager as superman flounders on the kryptonite of Kuper and Szymanski's 89 per cent.

One or two of history's finest do escape their censure, though: both Bill Shankly and Brian Clough emerge as marginal superheroes, the Robins to the money men's Batman. The rest, though, are brought out to be a very little kind of men. 'Most other managers simply do not matter very much, and do not last very long in the job,' the authors wrote in the first edition of *Why England Lose*. 'They appear to add so little value that it is tempting to think that they could be replaced by their secretaries, or their chairmen, or by stuffed teddy bears, without the club's league position changing. Even Manchester United's manager, Alex Ferguson, who has won more prizes than anyone else in the history of football, has probably performed only about as well as the manager of the world's richest club should.'[12]

By the second print run of their book, their stance had softened. With good reason: once this 89 per cent figure is placed in context it is not nearly such an indictment. The figure is derived by averaging a club's wage bill relative to its competitors' over the course of a decade and then seeing how closely it tracks a club's league position, also averaged over the decade. So we took a decade's worth of Premier League wage and league-rank data from Deloitte's annual financial reports – only we fast-forwarded to the most recent decade to cover the 2001/02–2010/11 period. A picture of consistency emerged (Figure 49).

Wages and league position go hand-in-hand, and the connection is tight: the higher the club's wages relative to the league

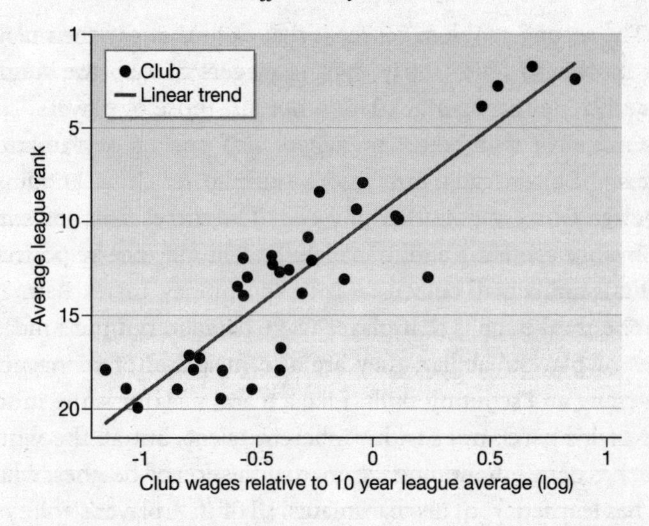

Figure 49 Wages and league position, Premier League, 2001/02–2010/11 (whole period)
Data source: *Deloitte Annual Review of Football Finance*, various years.

average over the course of the decade, the higher up the table the club finished.

For the past decade in the Premier League, wages explain 81 per cent of the variation in average final position. That's a little lower than Kuper and Szymanski's figure, but that could be explained by using different years, or because they chose to include the Championship. The message is clear: if you pay better you do better.

Before the anti-cultists scent victory and wheel out the guillotine, though, there are several problems. First, according to our calculations, the scraps for managers to influence seem to be nearer 19 per cent. Still not much, maybe, but at least better than the relatively paltry 11 per cent offered by Kuper and Szymanski.

The second problem we see is that clubs that pay their players more also tend to pay their managers more – the wages data include managerial salaries, not just those of players – so it's plausible that better managers also end up with better clubs.[13] Bolton can't hire José Mourinho or Guus Hiddink; Chelsea won't appoint Sammy Lee. The correlation between club wage spending and managerial talent may not be perfect, but it is unlikely to be zero.

The third issue is that player wage data are not pure measures of players' ability; they are also measures of managers' coaching and scouting skills. Like a team, a player is the product of his parts: not just his inherent talent, but all the work that has gone into honing it from a succession of coaches, what he has learned from his teammates, all of it. A player's ability – a key factor in determining how much money he earns – includes the input of his current coach, and all the coaches he's ever been with. Add in the fact that it may well have been his current manager who plucked him off the transfer market, and suddenly the role of the *mister* is not quite so irrelevant.

So credit for the very strong correlation between wage spending and league finish must be shared with the managers, who are an intrinsic part of finding and developing the best players.[14] Moreover when the decade-long numbers are sliced into individual seasons, the power of the paymaster starts to weaken.[15] According to Sue Bridgewater from Warwick University, the average tenure of football managers in England has dropped from over three years to less than a year and a half over the past two decades.[16] When clubs have half a dozen managers over a period of ten years, it is probably more useful to see what the connection is between wages and league position for this year and the next rather than over an entire decade. Managers are concerned with the here and now.

And when we look at the year-on-year figures (Figure 50), a very different picture emerges.

There is still a strong positive correlation between paying top wages and finishing high in the Premier League, but the rule is not that simple. There are plenty of clubs below the regression line – these are clubs that do worse than their wage bill – and above – the clubs that outperform their salary tab in any given season. Over the course of a single campaign, the amount of variation in league position explained by relative wages drops from 81 per cent to 59 per cent. There is a lot more room for immediate managerial influence; the Great Men are no longer condemned to fighting over the scraps; accountants, push the guillotine back out of the square.

Football is a game of balance, of light and dark. Whether it is in choosing between attack and defence, winning or not

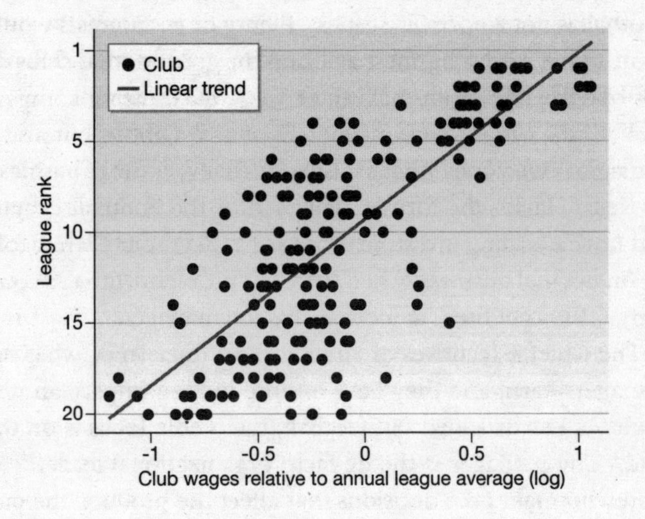

Figure 50 Wages and league position, Premier League, 2001/02–2010/11 (year on year)

losing or prioritizing whether to keep the ball or not to give it back, it is a sport defined by choices. Much of it – as much as half – is decided by fortune, cruel or kind, but there is a substantial part of it that is not, that is influenced by human endeavour. Some of that is players' skill. Some of it is managerial ability. It is these men who make the choices that determine a club's fate, or at least determine that part of a club's fate which cannot be attributed to luck.

Football is decided by fine margins. It is here that the manager comes into his own. To find out how much difference he really makes, though, we must look at his counterparts in other sectors of the economy.

Football Managers Are Like FT Global 500 CEOs

Football is not a normal business. Plenty of economists would argue that it's not a business at all, on the grounds that clubs do not operate like profit-making or value-maximizing companies.[17] Clubs usually have a Chief Executive Officer, but also a manager – who does what a CEO does in every other business. As Keith Harris, the former chairman of the Football League and now a leading investment banker at Seymour Pierce, told us: 'In normal business, when there is a problem with the company, CEOs get fired; in football, it's the manager.'

The Chief Executives of clubs make a fraction of what the managers earn, and they have a more limited impact on revenue. As Harris added: 'In football, the whole focus is on the field.' The manager is the de facto organizational leader: the man who makes the decisions that affect the product, the guy who hires and fires, and who is the public face of the club. He is the CEO in all but name.

As they damn this particular CEO with less-than-faint praise, Kuper and Szymanski note, 'The general obsession with managers is a version of the Great Man Theory of History. Academic historians, incidentally, binned this theory decades ago.' Recently, though, business-school professors and economists have pulled what might be nowadays termed the 'Great Person Theory' out of the bin, wondering who was so wasteful as to toss away a perfectly good hypothesis. Ignoring the ideology and the politics, they have examined creatively and thoroughly what the numbers say about whether leadership matters. They have proved that CEOs are of considerable importance.

A landmark study in the early 1970s dissected the performance of 200 large American companies and found that 30 per cent of a company's profitability was due to the industry it participated in, 23 per cent to its own history and structure, 14.5 per cent to the CEO, and the remainder to a variety of smaller factors. (The most technically advanced study of the influence of the manager on the fitness of a football club arrived at a figure of 15 per cent, almost identical to the business CEO figure.[18])

The critics are partially correct: a business's systems, structures and institutions are the main drivers of performance. If Steve Jobs had decided to go into typewriters and redesign that obsolete machine, Apple may not be so world renowned. If Sir Alex Ferguson had gone into water polo, he might never have floated out of obscurity. Leadership is not as important as the industry you're in or the organization you're part of.

But then Manchester United as a historical football institution is never going to change. These qualities are constants, and they provide both a maximum and a minimum limit for your company's success. Retailers will always be retailers. So long as they're trading, Boots and Tesco and Sainsbury's will always have their histories. There is only a certain amount of their financial performance that can change. Alan Thomas of the University of Manchester found that when you take out the fixed, unchanging elements of a company's performance, the impact of leadership on what remains rises to a range of between 60 and 75 per cent.[19] Of the things that actually can affect a club in the short and medium term, leadership might be the most significant.

Whether the business is retailing or football, in any industry where the margins between success and failure are small, even a percentage point or two of difference can be of considerable import, whether it's a 0.05 per cent upturn on a turnover of £1 billion, or five more points in the table.

That is the attitude taken by the management of the Tampa Bay Rays baseball franchise in US Major League Baseball, as described by Jonah Keri in his book *The Extra 2%*. The Rays are fast becoming the poster child of sports analytics. The majority owner, President and Head of Operations had all been trained on Wall Street, and they run the club with the aim of looking for 'positive arbitrage' possibilities.

Instead of searching for the figurative home run, Rays management have been looking for a continuous stream of small advantages and margins. Keri quotes the owner Stuart Steinberg: 'We've worked hard to get that extra 2%, that 52–48 edge.'[20] That edge has brought them to the play-offs in four of the last six years despite a total wage bill that was the fourth lowest in Major League Baseball, way down on the sums paid out by the New York Yankees or the Boston Red Sox. In football terms, it's Sunderland reaching the Champions League knock-out stages three times in half a decade.

The Rays also have, in Joe Maddon, arguably the best manager in baseball; the team's Wall Street-trained executives knew the importance of hiring a skilled leader. General Manager Andrew Friedman says, 'When we sat down with Joe and went through the interview process, it was apparent that his thought process was similar to ours in a lot of respects, in terms of being very inquisitive and trying to view things differently than maybe is conventional.'[21]

That bankers turned baseball men are willing to invest in leadership reflects the new interest of economists in the subject. To many in the field, the Great Person Theory remains moribund. A firm is nothing but a production function turning capital and labour into output, with labour's flow regulated by wages and incentives.[22] This was where Kuper and Szymanski started: if you want a striker to score more goals or a defender to make more tackles, pay them more or buy a better one.[23] The paymaster is at the wheel; the manager is a teddy bear. Such a view – thanks to research like the Koret workers study – is now seen as limited. Other factors – teamwork, leadership – are now afforded increasing importance.

To test this, a few economists have drifted into the rather gloomy field of death. Good CEOs and bad CEOs die with equal frequency: everybody's time will come, regardless of how much profit they've turned that quarter. Nothing could be more amenable to a sound economic study. Using what they termed 'a horrid empirical strategy' of cataloguing the deaths of CEOs and of their immediate family members, a trio of economists studied the finances of 75,000 Danish companies, and compared the financial results of companies touched by death to those whose leaders avoided the Reaper's scythe.[24]

This has happened in football, notably to Jock Stein, the legendary Celtic manager who lifted the European Cup with the 1967 Lisbon Lions. Jock, who was not Catholic, had shown himself to be acquainted with matters mathematical when he denied that his hiring had broken any barrier because fully '25 per cent of our managers have been Protestant', there being with his hiring four managers total in the club's history. Jock suffered a heart attack on the sidelines of a World Cup qualifying game in 1985 and later died. Scotland did still qualify for the 1986 World Cup but the only hint of success was a goalless draw with a Uruguayan team reduced to ten men. This was a far poorer performance than Scotland had managed under Stein in 1982. But was his absence the cause of it? We think it may have played a role, though we can't draw any definitive conclusion from the case of one leader and one club.

The ghoulish Danish economists, though, had a database of more than 1,000 CEOs who had died in office and 1,035 Danish companies that felt the loss and continued operations. If death takes a CEO and if leadership matters, profitability should decline both because there may be a vacuum until a new leader is in place and because that new CEO has, by definition, less experience running the firm than the dearly departed.

In addition, even though the king himself may endure, a death in his family may impact his performance. Will profits suffer if a bereaved CEO is less attentive? The anti-cultists would say no. And, in this case and that of the CEO's own death, they would be wrong.

Our trio of gloomy economists did indeed discover that CEOs matter statistically and economically: the demise of a CEO dropped profitability for the next two years by 28 per cent, and a death in his or her family contracted profits by 16 per cent. Leaders must matter because their absence or their inattentiveness causes performance to plummet.

Interestingly, the death of a director on the board caused no contraction or impairment in business performance, indicating that it wasn't the oversight and broad strategic functions of the

CEO that were missed but rather his or her operational activities. It's the hands-on actions of leaders that are most critical.

The implications of these data are clear. Money and wages are great. More money is better, in football as in real business. But leaders matter. They really do. Managers are far more than just stuffed teddy bears; rather, they are modern major generals only partially constrained by history and structures.

Having a good man in charge of your club will lead to improved results, a better league finish. Getting the appointment wrong, however, will mean that the portion of football that is not determined by fortune passes your club by. Results will dip, players will grow disillusioned, weak links will multiply, and the fans will drift away.

Critics would say that managers are responsible for only 15 per cent of their club's fortunes. That should suggest the anti-cultists, rather than the managerial loyalists, are correct. But football is a sport of the finest margins and 15 per cent is more than enough to be the difference between victory and defeat, between glory and failure. How, then, can a club be sure it has got the right man with information animal and acquaintance mathematical to lead it into fights historical? If managers matter, what makes a good one?

11.

The Young Prince

This is not a one-man show.
Maybe you can call me the Group One.

André Villas-Boas

If ever there was a season that encapsulated the varying impact a bad manager and a good manager can have on a football club, it must be the roller-coaster experienced by Chelsea, vintage 2011 to 2012.

Ever since Roman Abramovich took charge there, Stamford Bridge has provided the backdrop for English football's most captivating costume drama. It has been home to some of the game's most enthralling characters, with a heady mix of heroism, villainy and scheming, with a few power plays thrown in for good measure. The cast has been so rich and the twists so jaw-dropping that Shakespeare would have been proud. In the Russian oligarch overseeing it all, there's even the perfect *deus ex machina* to resolve the knottiest of tangles.

The months between June 2011 and May 2012 were pretty impressive even by Chelsea's standards. The season started with Abramovich appointing as manager André Villas-Boas, the thirty-three-year-old coach of FC Porto. He had worked at Chelsea before, under his now estranged mentor José Mourinho, before embarking on his own managerial career. He had, like Mourinho, found himself at Porto, where, like Mourinho, he had achieved considerable success in very short order. He was hailed by no less an authority than Gabriele Marcotti, the respected European sports journalist, as 'Portugal's boy genius'.[1] He seemed a natural fit for Chelsea: one of the most promising coaches in the world, with an added soupçon of the homecoming of the prodigal son.

Villas-Boas promised exciting, attacking football, overhauling Chelsea's reputation for dour, mechanized, heartless efficiency. He would provide the scintillating style that Abramovich longed for. And he would do it all with far more humility than Mourinho, the self-proclaimed Special One, could ever have managed.

Except it didn't work out like that. After eight and a half months, with Chelsea exiled from the top four of the Premier League and after a 3–1 first leg defeat in the last sixteen of the Champions League against Napoli, Abramovich felt he had no option but to dismiss the boy prince. He had enjoyed just 256 days in charge. Richard Bevan, Chief Executive of the League Managers Association, said the next man to take over would find himself walking into 'hell'.[2]

Not quite. Roberto Di Matteo, Villas-Boas's assistant and a former player at the club, took charge for the final three months of the season. He led the very same players whom Villas-Boas had so alienated to victory in the FA Cup against Liverpool, guided them past Napoli, Benfica and the mighty Barcelona to reach the Champions League final, and there, in Munich, against Bayern, he won the trophy that Abramovich had craved for a decade.

Bad managers fail, good managers succeed. The only changing factor in Chelsea's season was the identity of the man in the dugout. Everything else was constant. History will record that Chelsea made a mistake in appointing Villas-Boas and only drew back on course when they sacked him.

Knowing what we now know, though, about fortune's role in football, about the fleeting nature of possession and the importance of weak links, maybe that assessment is too harsh or insufficiently analytical. There is only a certain amount of influence – perhaps just 15 per cent – a manager can have. Thankfully, just as the numbers can help a manager choose the right path, they can also help a club choose the right manager. Not only that, they can guide them as to how to make sure their choices are successful.

Andrew Friedman, general manager of the Tampa Bay Rays, a club we've seen are at the forefront of analytics, says his club always 'post-mortem' decisions at a later date. 'We keep copious notes on the variables we knew, everything we knew going in,' he says. 'Then we go back and look at it to review the process. It's something we're continuing to refine and will be in perpetuity. I hope to never get to the point where we're content, or we feel great about everything and go into autopilot mode.'[3]

Perhaps it is time to apply the same logic to the appointment of football managers. And what better case study, what better post-mortem, than the boy wonder who flew too close to the sun?

Crowning the Young Prince

The two greatest sources of concern that arose when Abramovich chose to pay Porto around £13 million to hire Villas-Boas were his lack of a high-profile playing career and his lack of managerial experience.

Premier League clubs like their managers to have been players. Between 1994 and 2007, data show that more than half the managers in the top flight had played the game to a high enough level to represent their national team. In League Two over the same period, it was one in seven.

This obsession with former players is often seen as a weakness. As Arrigo Sacchi famously noted when questioned about his qualifications, 'I never realized that in order to become a jockey you have to have been a horse first'; the best students may not make the best teachers. Many of the foremost managers of the current era – Mourinho, Wenger, Benítez – were either mediocre players or didn't play at all. Successful players are more likely to hark back to the methods that made their careers glorious, rather than adapting and innovating, as managers at all clubs must, since there is no permanently winning formula.

Clubs appointing managers because of the players that they were are seen as short-sighted, certainly by fans of an analytical bent, and even by some within the game. 'That's the culture of English football, and it has always been that way,' Andy Cale, Head of Player Development and Research at the Football Association, says. 'Clubs have always gone for famous ex-players who were seen as winners. In the last ten years some chairmen have become a bit more clever, but it's obvious that it will take time. In the meantime, however, this attitude in choosing managers has had disastrous effects. Just look at the number of sackings each year.'[4]

The numbers, though, tell a different story. Using twenty years of Premier League and Football League data, economists Sue Bridgewater, Larry Kahn and Amanda Goodall confirmed in 2009 that a manager who had played for his national side was generally more effective than someone who had never won a cap.[5] Managers who had been skilled players themselves were particularly effective when they were in charge of teams of lower-paid and lesser-talented players.

Many in football believe that top players struggle as managers because they cannot teach what came instinctively to them. They cannot, the theory goes, communicate their knowledge. Not so, according to the data. 'Look, if you were a good player you can teach things others cannot,' Fabio Capello asserts. 'There are elements of technique, of timing, of coordination which I don't think you can understand if you never played the game at a certain level.'[6]

In the case of Villas-Boas this may not have been an issue. At Chelsea, he had some of the finest players in the world at his disposal, even if it did not always look like it. There was not much that these players could be taught, even by a Johan Cruyff or a Franz Beckenbauer. Old dogs, like Ashley Cole, already know all the tricks.

Villas-Boas's lack of managerial experience may have been more significant. The numbers show that, on average, a Premier League manager has managed for nine years. Villas-Boas had done just two: one at Académica de Coimbra, one at Porto.

Bridgewater and her co-authors found, unquestionably, that 'more experienced managers bring more highly skilled players closer to their potential'.[7] In the case of Chelsea, the effect of a relative newcomer like Villas-Boas on the performance of superstars would have been considerably different from that of a manager with greater experience. The numbers show that Chelsea could well have sacrificed two or three places in the league table by replacing the secure Carlo Ancelotti with the young, exciting Villas-Boas rather than the wise Guus Hiddink, the other contender. Frank Lampard, John Terry and the rest would have responded better to a wrinkled old hand than a manicured young one.

All this information was available to Chelsea when they were assessing their managerial appointment and the trade-off between Villas-Boas and Hiddink or someone like him. Only insiders at the club will know whether that decision-making process in summer 2011 adhered to the Tampa Bay Rays' rule of considering all known variables before coming to a conclusion, but the evidence suggests it did not.

Abramovich and his executives knew the talent Manchester City and Manchester United possessed in their squads and the calibre of their managers. They should have forecast that even with Ancelotti, Hiddink, or another very experienced manager at the helm, Chelsea were likely to finish no better than third or fourth in the Premier League during the 2011/12 season. Under a manager like Villas-Boas, that expectation should have slipped two places, to fifth or sixth: where Chelsea were when they sacked the Young Prince, and where they finished in the league. This cost could have been anticipated based on the 2009 research, and yet it looks like it took Chelsea by surprise. To the Tampa Bay Rays or to any well-run analytical organization, this particular surprise indicates a suboptimal decision. Abramovich could have known that appointing Villas-Boas would lead to a drop in the table, and therefore, he *should* have known.

Assessing His Rule

There can be no greater justification of Abramovich's risky decision to appoint Villas-Boas than the remarks of a General Electric executive speaking of his company's approach to manager development. 'Bet,' the executive proclaims, 'on the natural athletes, the ones with the strongest intrinsic skills.

Don't be afraid to promote stars without specifically relevant experience, seemingly over their heads.'[8]

This mindset of looking for 'natural athletes' dominates the global competition that the consulting firm McKinsey termed the 'war for talent'.[9] Organizations as diverse as Google, General Electric, Barclays, Bain and Oxford University all compete for human capital, usually following Florentino Pérez's prescription at Real Madrid: collect as much expensive individual talent as you can afford.

Football is no different. The GE executive's words mirror the beliefs that fuel much of football's outdated recruitment strategy and that Thomas Carlyle, the Great Man advocate from last chapter, would verify: talent is innate, given as a gift from the supreme being; it can be identified from afar and at an early age; and thirdly, talent is wholly possessed by the person, so that it can be bought and sold and moved around without friction.

Sadly, all of this is bunk.

Talent, whether it is musical or athletic, is not innate. It is nurtured. That is true for both Mozart and Tiger Woods; as Geoff Colvin argues in *Talent Is Overrated*, neither of these were prodigies born into the world tinkling the ivories or driving down the fairway; they were both developed by (pushy?) parents over the course of hours and hours of practice. Likewise, Pelé, Maradona and Messi were not born with a ball at their feet.[10]

This has been established by an ingenious study. To see if it is nature or nurture that produces exceptional performers, a group of British psychologists tracked a group of 250 young musicians of varying abilities.[11] Talent, they found, does not blaze like a beacon. There were, they recorded, 'little or no differences between high-achieving young musicians and others in the level and age of incidence of very early signs of musical behaviour or interests that have often been supposed to be signs of exceptional "talent"'.

Secondly, the psychologists found that there is a strong correlation between practice and achievement: 'High achievers practise the most, moderate achievers practise a moderate amount, and low achievers practise hardly at all.' Talent is a function of work. This has led to the 10,000-hour rule of thumb, made famous by Malcolm Gladwell in his book *Outliers* as the amount of time needed to master any skill.

What does this mean for footballers or managers? You are not born with a gift; you must work at it. What, eventually, may separate one hard-working youngster from another is that he or she is labelled as 'one to watch'. And we've reviewed enough psychology here to know that this label may not be a rational judgment or scientific assessment.

This is all true of Villas-Boas. He was not born a brilliant manager; he worked at it, first through hours of playing football-management simulation games and later by assessing FC Porto games for a school project. He was dedicated, but he was also lucky: an insightful look at the young Villas-Boas's life by Duncan White of the *Sunday Telegraph* revealed that he happened to live in the same apartment block as Sir Bobby Robson, then manager of Porto. Villas-Boas made sure he bumped into the Englishman and asked him why he was not playing Domingos, the centre forward. Robson asked him to compile a report on the issue. Robson liked what he saw, so he tasked Villas-Boas with more research, taking him to training occasionally.[12]

Villas-Boas received a classic apprenticeship. Our blacksmithing forefathers may have had a better, and more modern, understanding of talent than we do: it's less about the selection – any willing hard worker will do – and more about the training – there are right ways and wrong ways of doing things and there is specific knowledge you have to have. The apprenticed is a quite distinct person from the anointed.

When Chelsea appointed Villas-Boas he was seen as a blazing beacon of ability. When he was sacked he was an ash-heap of incompetence. The truth lay somewhere in the middle. Chelsea's evaluation of Villas-Boas's skills should not have ended once he signed a contract. Because talent can't be judged very accurately from afar, once it comes close you want to see what you have.

Chelsea, it is fair to say, were probably not assessing Villas-Boas's talent levels during that season. It's difficult to gauge, after all. Clubs cannot put a season on hold while they conduct an experiment.

That is not to say there are no data. Take the first sacking of that 2011/12 season, the dismissal of Steve Bruce (described by Barney Ronay as resembling a 'family butcher with a secret') by Sunderland. Ellis Short, the club's owner, replaced him with Martin O'Neill, a man who ticks all the boxes for a Premier League coach: vastly experienced as a manager and as a player.

We would have advised Short to welcome his new man not by giving him a vast transfer kitty for the January sales or by shipping out underperforming players, but by doing nothing. That's the only way to get clean data.

This is not as wilfully negligent as it sounds. The correlation between transfer spending and on-pitch performance is reasonably strong. In January 2012 Sunderland's squad was valued at £95 million, placing them tenth in the Premier League money table. In reality, they were seventeenth, many points below where they should have been, thanks to chance and the family butcher. Spending an extra £20 million to bolster the squad would have put them . . . tenth in the Premier League money table. They would have overtaken nobody. It would have been money down the drain, and it would have given Short a way of knowing whether his new manager had made a difference.

Short was obviously listening. Sunderland bought nobody and sold two fringe players. The conditions existed to assess O'Neill's effect. Sunderland finished the season in thirteenth place, much nearer where a squad of that value should have been. His owner now had reason to believe the Northern Irishman had done a better job than Steve Bruce.

But where Sunderland had an answer, Chelsea did not. They have no idea whether Villas-Boas was any good or not, because the conditions did not exist to assess his performance. Ideally, when they hired him, they should have just given him Carlo Ancelotti's squad of the previous season to work with. If the end result was first place, then Villas-Boas clearly had more talent than his predecessor.

This is unlikely, and it is also slightly unfair: the players would be older, and fortune may not affect the two managers equally. So Chelsea should have done all they could to keep the squad as stable as possible in order to clearly assess the Young Prince. Players should only have been signed if their extra value over the player they replaced was such that it outweighed muddying the data about the manager. That, it is safe to say, did not happen. Chelsea spent and sold to the tune of £107.4 million in the summer of 2011; their fourth most lavish summer spree under Abramovich. Eighteen players came and went, so change was loaded upon change.

Maybe Villas-Boas wanted some of these deals done (knowing Chelsea's reputation, it's certain that Abramovich wanted most of them), despite his public profession that he was 'more than happy with the actual squad'. He wanted a younger squad, this much is well known, and the club wanted to see a different type of player brought in. That's all well and good. But because doing so rendered the data on Villas-Boas useless, it means assessing his ability, his performance, is almost impossible. He could be either the charlatan of March 2012 or the superstar of June 2011. Chelsea will never know.

Power Behind the Throne

Six months before Villas-Boas arrived at Stamford Bridge, Abramovich's billions brought an even more glamorous prodigy to west London. Fernando Torres arrived from Liverpool for a British record fee of £50 million and was hailed as the blond bombshell who would transform workmanlike, automated Chelsea into a team of beauty and élan. It did not quite work. In his first eighteen months at the club, the previously prolific striker scored just twelve goals; in his first full season, 2011 to 2012, he did not score a Premier League goal between the end of September and the very final day of March.

As Torres's form nosedived so did his body language. He seemed disinterested on the pitch, pouting at his own failings. The problem was attributed to his failure to adapt to his new teammates, to Chelsea's style, to playing in the shadow of Didier Drogba, and even, according to Terry Venables, Torres's own lack of work ethic. 'I have known of players taking time to settle but not to this extent,' the former England manager wrote. 'He needs to start working harder in training but stop trying so hard in matches. There is no short cut.'[13]

To Venables, the reasons behind Torres's slump were irrelevant. To his mind, everything can be cured by getting out there on the training pitch and doing a bit more shooting practice.

Overseeing that during that long, barren spell would have been Villas-Boas. We suspect he may have known exactly what the Spaniard was going through, and would have been unable to help him for the very same reasons that Torres was unable to hit the back of the net.

The shared plight of the Iberians reflects the third mistaken aspect of the war for talent: the idea that ability is encapsulated within an individual, and so can easily be moved around, bought and sold. Venables's statement that he has 'known of players taking time to settle' is more astute than he may realize.

For many years talent was largely frozen in place in football. Oleh Blokhin, one of the finest players ever raised in the Soviet Union, won eight titles with his Dynamo Kiev side in the 1970s and 80s, scoring 211 goals in the process. Any top team in Europe would have loved to sign him, but could not, because the Soviet authorities would not let him leave the country. Only when he was past his prime did Blokhin move abroad, in 1988.

Two years later Boris Groysberg moved with his family from the Soviet Union to the United States, where he rose to become a professor at Harvard Business School, specializing in the portability of performance that Blokhin – now coach of the Ukraine – never had to contend with. Groysberg wrote a book – *Chasing Stars: The Myth of Talent and the Portability of Performance* – about the ultimate free agents of the corporate world: equity analysts on Wall Street.

These analysts are experts on a particular sphere – retail or pharmaceuticals or sport – and they spend their time writing reports evaluating the prospects of companies in their chosen industry and making predictions as to their likely performance. These predictions, in turn, provide quite a neat way of analysing the performances of the analysts themselves, to see if they are worth the multimillion-dollar salaries the very best of them attract. If their predictions are accurate, they are worth every penny they earn.

All the outward signs suggest these analysts are just plug-and-play: take them to a different bank, give them their computer and their files, and off they go researching their companies. As one director of research at a bank says about analysts' portability, 'I mean, you've got it when you're here and you've got it when you're there. The client doesn't change. You need your Rolodex and your files, and you're in business.'[14]

Groysberg, though, found that this is not the case at all. With his research colleagues, he gathered numbers for all the equity analysts in the United States who were ranked by the industry magazine *Institutional Investor* – which publishes a list of all the top analysts every year – and then identified the 366 of those listed who had changed banks between 1988 and 1996.

Now, *Institutional Investor* ranks only 3 per cent of all the many thousands of analysts in the US every year. This is the cream of the crop, the all-stars. And what Groysberg found was stunning: if ranked analysts stayed put at their bank, there was an 89.4 per cent chance they would be ranked again the next year. That slipped to 69.4 per cent if they moved to a rival company. The analyst ranked top in any given year had a 10.6 per cent chance of retaining the highest place if he or she stayed put, while those who moved saw their chances drop to 5.6 per cent.

There were longer-term performance effects as well: those first-team all-stars who remained with the same bank had a likelihood of 54.3 per cent of returning to the top ranking at least once over the succeeding five years; superstars who transferred had a likelihood of only 39 per cent.

The parallel with football is clear. The Sporting Directors at Paris Saint-Germain and Bayern Munich assume that a player or a manager is plug-and-play: drop them into or in charge of a team and off they go. Talent, they think, is portable. To paraphrase our bank director: 'You need your boots and your shin pads, or your clipboard and whistle, and you're in business.' The numbers say it is not nearly as simple as that.

Groysberg has proven that his principle, unlike free agents themselves, transfers easily to other fields, even sport. In American football the punter has just one job. When teams (incorrectly, remember) choose to punt on fourth down, he receives the ball from the snap and is tasked with booting it as far down the field as he can. Wide receivers, on the other hand, are part of a unit: they run in patterns, coordinating with the quarterback, adjusting to the opposition's defence on the fly.

Groysberg found that the performance of free-agent punters, who are literally standalone players, does not decline after switching teams, but the key statistics of the receivers dropped for a full year after signing with a new team. In football there are no true standalone players – goalkeepers come closest, but even they must interact with their defence – so we must expect a period of adjustment.

In *Chasing Stars*, Groysberg recommends that, to minimize these effects, companies or clubs do their best to promote from within.[15] Where they cannot, they must have a systematic plan to add only those outsiders who fit the culture and who are then assimilated deliberately and carefully into the team. He writes, 'Hire with care but integrate *deliberately* and *fast*.'

This does not seem to occur to football clubs. Until very recently, very few of them employed specialists to help players settle in off the field, often leaving them to find their own homes and trail round schools to sort out the children's education. At

Chelsea, too, there was a culture of leaving new players to get on with it.

'I plunged into problems linked to my situation as an expatriate,' Drogba, Torres's predecessor, wrote in his autobiography. 'Chelsea didn't necessarily help me. We sometimes laughed about it with Gallas, Makélelé, Kezman, Geremi. "You too, you're still living in a hotel?" After all these worries, I didn't feel like integrating.'[16]

This can be helped by executing what is known in the corporate world as a 'lift-out': hiring a star and some of his teammates. Groysberg found that top analysts brought into a new company alongside several co-workers experienced no decline in performance. Chelsea did this in hiring Mourinho – four staff, including Villas-Boas, followed him from Porto, as well as two players – but did not repeat the trick when they appointed his young protégé seven years later. Villas-Boas brought just two staff with him. His assistant, Roberto Di Matteo, having left Chelsea's playing staff some nine years earlier, was to all intents also new, acclimatizing. No wonder Villas-Boas did not feel comfortable.

Nor did Torres, despite joining his former teammate Yossi Benayoun at Stamford Bridge and quickly being reunited with Raul Meireles, too. That is not enough. Mourinho had six familiar faces around him. Maybe he, and not his former protégé, should have referred to himself as the Group One.

Regicide: Dethroning Villas-Boas

The end, when it came, was dismal. Chelsea had written off the Premier League title as early as Christmas; they were on the cusp of elimination from the Champions League and had just been beaten by West Bromwich Albion (ironically, the club that had sacked Roberto Di Matteo a year earlier). Villas-Boas was facing mutiny inside the dressing room and had lost the faith of Abramovich, the key power broker. He had held emergency talks with his squad after two particularly bad results, at Goodison Park and the Hawthorns, and been denounced in no uncertain terms. He was finished. When he was sacked it was almost a kindness.

On the surface, the change of management worked a charm. Under Di Matteo, Chelsea took ten of an available fifteen points in the league, rescued their FA Cup campaign and, remarkably, smashed Napoli 4–1 at Stamford Bridge to put them on their way to Champions League glory. Getting rid of Villas-Boas, it was written, was the best thing Chelsea had ever done.

In the last five Premier League games under Villas-Boas, Chelsea had averaged a mere point per game; in the final eleven under Di Matteo, the club averaged 1.64 points. It seems glaringly obvious that Di Matteo's leadership was revitalizing. Or is it? One number suggests not: in twenty-seven league matches under Villas-Boas, Chelsea averaged 1.70 points per game, a better rate than Di Matteo's. Now, this could be because the squad switched their attention from domestic duties to foreign affairs as the Champions League final drew ever closer. Or it could be that their improvement was an illusion. They may simply have regressed to the mean.

This is the numerical equivalent of the physical idea that water will always find its own level: extreme numbers will usually be followed by medium ones; giants and the diminutive will have normal-sized children. You can't play the numbers game without a thorough understanding of regression to the mean.

The most statistical of team sports, baseball, is a great place to observe this phenomenon. Batters in baseball come to the plate to hit around four or five times every game, and a batting average is the percentage of hits they get out of all their plate appearances.

An adequate batter will have an average of .250 (25 per cent hits), a very good one will hit .300, and occasionally a player will have a great season and hit .350 or above.[17] Figure 51 shows what happens the following year for players who had batting averages of .350 and above in the major leagues after World War II.[18]

The extreme year is shown on the horizontal axis and the following year on the vertical: if the next year's average is lower, the point will appear below the slanting line. Regression to the mean is illustrated by the fact that almost all of the points are below the slant, and most are well below: extremes

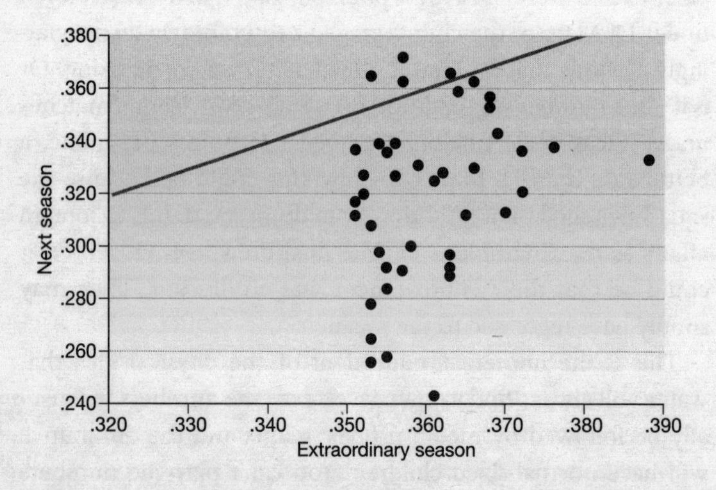

Figure 51 Regression to the mean in Major League batting averages, 1946–2002

are usually followed by intermediates, the extraordinary by the more ordinary.

Extreme events can be positive like a very high batting average, or they can be negative like a losing streak or a patch of poor form at a football club. This seems to have been a major part of Abramovich's decision to bite the bullet and sack the Young Prince; poor recent form, studies show, is always one significant factor when any manager is dismissed.

Macclesfield, one of the lesser lights of English football, provide an apposite example. In January 2012, just as Villas-Boas's own form was starting to tank, the Silkmen held Premier League Bolton to a 2–2 draw in the FA Cup third round. That was as good as it would get for manager Gary Simpson. The club lost every league game in the month of January, drew three and lost three in February, drew three and lost three in March, drew one and lost five in April. Macclesfield plunged to last in the table. Simpson clung to his job until March, when the club could no longer resist calls for his dismissal. At this point, it didn't matter. His replacement, Brian Horton, managed just two points from eight games and the club was relegated from the Football League.

Despite Horton's inability to reverse the trend, there is plenty of anecdotal evidence that the dismissal of a failing manager can revive the spirits and performance of a club. This has even been borne out by studies. Multiple investigations into sackings in leagues across the breadth of Europe have shown that clubs' performances form a trough with a downward slope leading to the sacking and then a climb back up as the points start to flow again.[19]

Figure 52 displays the numbers from a Dutch study, where 't' is the time of the sacking, minus or plus a number of matches.[20]

Performance at the typical club has declined drastically to

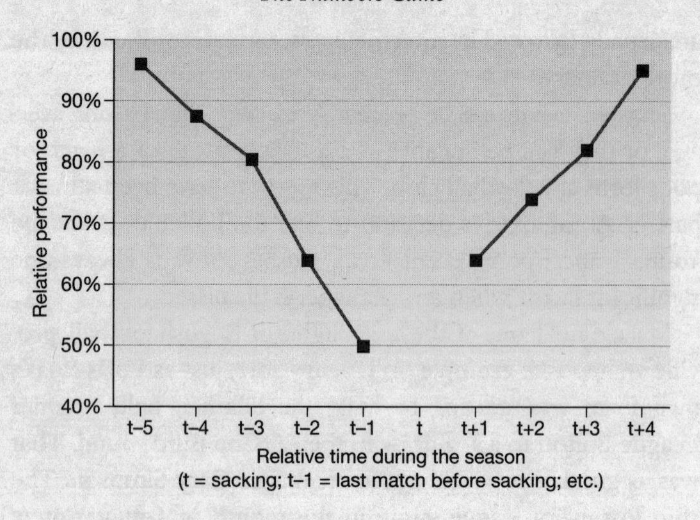

Figure 52 Club performance before and after managerial sacking, Eredivisie, 1986–2004

50 per cent of its potential in the week before the manager is sacked. By the fourth game under the new coach, performance is at 95 per cent of what it should be, the fans are sated and the boardroom is filled with the gentle sound of self-satisfaction. No doubt it was like this at Chelsea, too, as Di Matteo appeared to undo all of Villas-Boas's bad work. Abramovich – and other owners – just do what the numbers seem to demand of them.

This, sadly, is another beautiful hypothesis slain by ugly fact. Sackings do not improve club performances. Clubs simply regress to the mean.

To see if sacking the manager makes a difference, the author of the Dutch study, Bas ter Weel, searched within all the other non-sacking data from the eighteen seasons of Eredivisie results to identify a control group to compare to the sacking episodes. The control group consists of those spells (distrib-

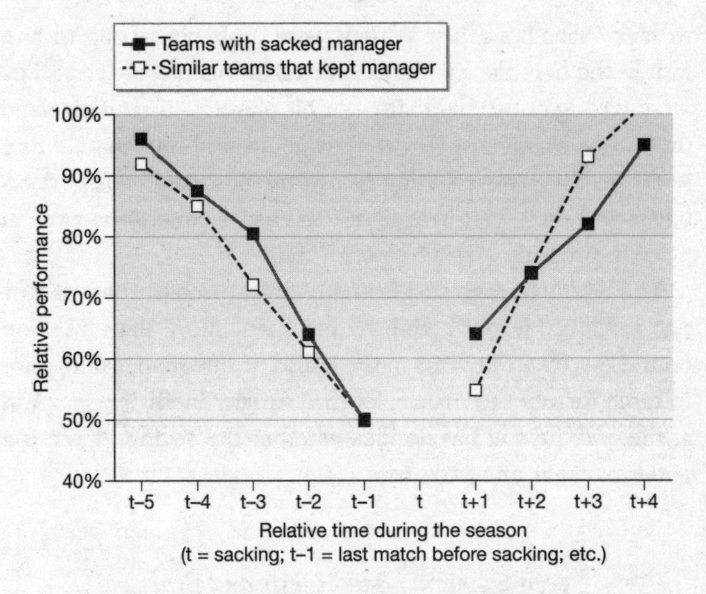

Figure 53 Club performance dips and recoveries with and without managerial sacking, Eredivisie, 1986–2004

uted statistically equally across all the clubs) in which a club's points per game average declined by 25 per cent or more over a four-game stretch, but they did *not* sack their managers. Ter Weel found 212 such cases. The results are shown in Figure 53.

Even without sacking the manager, the performance of the control group bounces back in the same fashion and at least as strongly as the performance of the clubs that fired their managers. An extraordinary period of poor performance is just that: extraordinary. It will auto-correct as players return from injury, shots stop hitting the post or fortune shines her light on you once more. The idea that sacking managers is a panacea for a team's ills is a placebo. It is an expensive illusion.

Chelsea did not get these decisions right. They were wrong

to sack Villas-Boas, just as they were probably wrong to hire him in the first place if they expected immediate success; they were wrong not to help him – or his players – acclimatize and they were wrong not to put a group around him to help him settle in. They were wrong to assume he could transport his talent and they were wrong not to ensure conditions were in place to properly assess his performance.

The whole farrago cost Roman Abramovich the best part of £30 million. He put a club he has spent more than a billion pounds on over ten years in the hands of one man, asking him to tame fortune, to master football's many inefficiencies. And at the end, he still has no idea whether the Young Prince is a good manager or a bad one.

Speed Cameras, Bad Habits and Practice with Your Head

Much to Jeremy Clarkson's chagrin, you can barely drive on a British road these days without seeing a Gatso, the widely loathed mounted yellow box which gauges drivers' speeds and captures miscreants on film, so allowing the police to issue them with tickets.

Gatsos were, originally, typically installed at places that had experienced a recent and unusual cluster of accidents. When the Department of Transport issued its evaluation of the speed camera scheme after four years, it claimed a 50 per cent reduction in fatalities and serious injuries thanks to the presence of the cameras. The Gatsos, despised as they were, had done their job. Except that they hadn't. Buried in the appendix of the report was a study by Liverpool University's Department of Engineering that concluded 'the presence of the camera was responsible for as little as a fifth of the reduction in casualties'.[21] Even without that yellow overseer, the number of accidents in black spots would have regressed to the mean. Extraordinary numbers of fatal accidents would be followed by more typical ones.

It is not just motorists and managers who regress to the mean. Everyone does. Players do, certainly. One game he will be outstanding, beating opponents, terrorizing defenders. The manager will shower him with praise. And yet the next week the same player will be ineffectual, unable to make an impact. Or the flipside: a defender is awful one Saturday, having a real Khizanishvili of a game, attracting a searing, puce-faced outburst of unmitigated rage from the manager, and then brilliant the next week.

Every manager will have had this experience, and no doubt it is why so many of them favour the stick over the carrot. If you're nice to players, praising them, patting them on the back, they grow complacent. Give them a kick up the backside and they perform much better.

This, too, is an illusion. The players' performances are simply righting themselves. Nobody can better explain why this is than Danny Kahneman, winner of the 2002 Nobel Prize in Economics, whose research has focused on the limits of rational decision-making:

> I had the most satisfying Eureka experience of my career while attempting to teach flight instructors that praise is more effective than punishment for promoting skill-learning. When I had finished my enthusiastic speech, one of the most seasoned instructors in the audience raised his hand and made his own short speech.
>
> He said: 'On many occasions I have praised flight cadets for clean execution of some aerobatic manoeuvre, and in general when they try it again, they do worse. On the other hand, I

have often screamed at cadets for bad execution, and in general they do better the next time.' This was a joyous moment, in which I understood an important truth about the world: because we tend to reward others when they do well and punish them when they do badly, and because there is regression to the mean, it is part of the human condition that we are statistically punished for rewarding others and rewarded for punishing them.[22]

Football management, it is safe to say, remains exposed to the same perverse contingency. Seamus Kelly and Ivan Waddington of the Centre for Sports Studies at University College Dublin conducted a series of structured interviews with twenty-two top-tier players and eighteen managers during the 2004/05 season.

They published their findings in a paper titled, 'Abuse, Intimidation and Violence as Aspects of Managerial Control in Professional Soccer in Britain and Ireland'. Their interviews are worth reading and reveal the depth and breadth of mean and nasty behaviour in football clubs. Because managers do not take into account the concept of regression to the mean, they simply tend to regress to being mean.

'[The manager] would hurl abuse at you all the time,' one player said. 'In front of other players, in the office, on your own, or in the office in front of the coach and staff. It brought out the best in me. It did bring out the best in me. But I know not all players could hack it, they just couldn't hack it . . . If you are a young lad just coming in, then a manager will just take their frustration out on young players generally.'[23]

Note that buried in the player's phrase 'bring out the best in me' is the regression to the mean – he was underperforming, he got yelled at, and then he played better. But that does not mean he got better *because* he was yelled at.

Managers must be confrontational, angry and passionate at times. But the better managers know when to deploy such tactics and when to use a different approach. Sir Alex Ferguson, famed inventor of the Hairdryer Treatment, probably used it less than we think: even when his side were losing 3–0 at Tottenham, he did not raise his voice. Musa Okwonga, in his book *Will You Manage?*, recounts Denis Irwin's version of what happened in the White Hart Lane dressing room at half-time:

'Ferguson didn't use the hairdryer. What he said, very calmly, was, "Obviously, you know that this is Spurs we're playing. In their minds, they've already won, they're in the pub after the game celebrating. Get a goal back at the beginning of the second half and they'll panic. That's the thing about Spurs. They've always played like that, and they always will." '[24]

This is much more effective than the obscenities or shattering teacups or fists many a manager flings at his underperforming players at half-time. United scored five after the break that afternoon. Their performance regressed to the mean, because Ferguson chose not to lose his rag: instead, he communicated his knowledge to his players. This is what good managers do.

Calmness and knowledge sharing should be the rule on the practice pitch as well. When observers kept logs of the training sessions of the two greatest coaches in American collegiate basketball – for men, UCLA's John Wooden, and for women, Tennessee's Pat Summitt – half their utterances were instructions, such as 'do some dribbling between shots'.[25]

More than 10 per cent of Coach Wooden's actions involved demonstrations of the correct or incorrect movement or both – showing the player the right way to do something.

Training sessions were fundamentally about instruction for the UCLA coach: 'I felt running a practice session was almost like teaching an English class. I knew a detailed plan was necessary in teaching English, but it took a while before I understood the same thing was necessary in sports. Otherwise, you waste an enormous amount of time, effort, and talent.'

This same principle applies in football. Talent is not given by God. It must be honed and nurtured, sculpted and shaped. Good managers, like Wooden, script every session, so that it has an aim, a result. They must also monitor their own learning.

'I kept notes with the specifics of every minute of every hour of every practice we ever had at UCLA,' said Wooden. 'When I planned a day's practice, I looked back to see what we'd done on the corresponding day the previous year and the year before that.'

Some managers in the modern Premier League are equally meticulous; Mourinho, certainly, and we suspect Villas-Boas. Others, less so. 'When we trained, if someone got a clear shot at goal then he [the manager] would stop training and make us run for twenty minutes,' one defender told Kelly and Waddington. '[For] any mistakes at all. We were terrified to make mistakes. It was entirely based on fear. We were scared. After matches that we lost, he would have us in at six in the morning running.'[26]

This kind of training is undoubtedly happening less frequently in elite football clubs, and it should continue to diminish. The new model should be based on the insight that has now become identified through the modern theory of talent as deliberate practice. As the violinist Nathan Milstein wrote: 'Practice as much as you feel you can accomplish with concentration. Once when I became concerned because others around me practiced all day long, I asked [my mentor] Professor Auer how many hours I should practice, and he said, "It really doesn't matter how long. If you practice with your fingers, no amount is enough. If you practice with your head, two hours is plenty." '[27]

The Model of a Modern Manager

The manager is far from an irrelevance. Yes, fortune is of huge significance in football. It accounts for around half of what we see on the pitch and in the final league table. Yes, money is a factor. But there is much more to this game, to the pursuit of the goal. There is a world of styles to choose from, a huge amount of factors to consider, and the very best generals will use all of the information they have to get the best out of the resources at their disposal. They will think laterally and innovatively, and they will find ways of changing the game for their own benefit. They do have an impact.

And yet many clubs do not seem to have any idea what makes a good manager. They lack the conditions to examine the current incumbent's abilities, and they seem to fall into the trap too often of replacing a failing coach with the flavour of the month, or with a household name with a track record of mediocrity. They react too quickly when things are not working out, and they do not do all they can to help their managers succeed (this goes for players, too).

But just as clubs must learn to hone their appointment process and to be patient with the man they have got – all the while assessing his ability – managers must strive to be the best they can. They must bite their tongues, resorting to anger only when absolutely necessary. They must monitor their own habits, share their wisdom and not be afraid to challenge convention. They have to do what they can with their resources. They must not fear going for it on fourth down. They must be aware that they owe a debt to the collective. If they do that, if they hold their nerve and stick to their beliefs, they can thrive.

The Young Prince, Villas-Boas, is ageing fast and is not thriving, having already experienced another cycle of managing, hired by Tottenham in July 2012 and then leaving 'by mutual consent' in December 2013. The good news for AVB is that he made it longer than a single year with a club and, with respect to percentage of matches won, he is third of all-time among Tottenham's managers, with a percentage of 55 per cent, greater than both Harry Redknapp's 49.5 per cent and Bill Nicholson's 49 per cent. However, the bad news is that he was denied the time and stability to hone his craft, and he replicated the troubles he had with experienced, talented veterans at Chelsea by falling out with Emmanuel Adebayor, the Spurs striker, who confronted AVB about his ideas and tactics in front of the rest of the team, and whom the still callow manager banished to the bench and to training with the youth squad. At this point, even the ever self-confident Ageing Prince has to have some doubts about his managerial potential.

Maybe managers such as AVB are scrapping at the margins. But football is a sport of rare events, a game of rare beauty. It is decided by the margins. It is at the margins where games are won and lost, where history is made and reputations forged, where light and dark meet.

After the Match
Is Before the Match

12.

Life During the Reformation

Football is a game you play with your brain.

Johan Cruyff

Sacred cows make the best hamburger.

Mark Twain

November 2011 saw the UK's first Sports Analytics conference, styled after the MIT Sloan conference in Boston. Held at Manchester University's Business School, the gathering was a relatively small, less glamorous affair than its American cousin. There were around 150 in attendance, drawn largely from football and rugby clubs, with a couple of Olympic sports thrown in. There were scouts, performance analysts, consultants and executives, but there were also a few managers and owners.

Phil Clarke, a star of the all-conquering Wigan Rugby League team of the 1990s and one-time Great Britain captain, is behind the endeavour, together with his brother Andy, formerly a fitness coach at Liverpool. The Clarke brothers, with a growing number of others, have recognized that sport is changing, and they have realized that with change comes opportunity. They have a company called The Sports Office, which helps clubs organize their in-house performance, administration, scheduling, medical, training, conditioning and scouting data.

The football data business is becoming competitive; similar companies are sprouting up around the world of professional sports. There's a need and a hunger for information – or rather, how to cope with information – but few have a plan for how to use it. One of the best-received presentations at the conference was by Tesco's Retailing Director, Andrew Higginson, on how Tesco mined its data to become number one. Analysts from football clubs listened intently – they all want to be number one – and as we have tried to explain in the chapters above, every little helps.

What we know already can give us a hint as to where the analytics reformation is heading over the next decade. This is our forecast for the journey that football and all those in it are about to take in the next ten years. We can never be certain, and some of our predictions will be wrong, but they are based on the best research and information that we have now.

Forecast 1

The biggest analytical breakthroughs will not occur at Manchester United, Manchester City, Real Madrid, Barcelona or any of the twenty richest clubs listed in Deloitte's Football Money League.

To get the delegates in the mood, the Manchester conference started with a recorded greeting from the patron saint of sports analytics, Billy Beane – the now famous Oakland A's general manager who turned baseball on its head by using numbers instead of gut and tradition to build a team.

Beane is a celebrity, thanks to Hollywood, but the truth is that he is just the latest in a long line of analytics innovators in sport. Charles Reep learned first-hand that football has always been resistant to change, be it the passing game, the idea that football could be played both for fun and for money, the 4–4–2 formation, or the notion that long balls can be quite effective under the right circumstances.

Those innovations that have prevailed have tended to be the ones that produce winning football: the passing game, *catenaccio*-style defence, the flat back four. There have always been innovators, men who don't mind thinking about what makes football work better. They are unified by a willingness to try the new, untested, perhaps slightly unusual; they are the men who go for it on fourth down. Some will succeed, and see their adaptations aped, and others will fail. There will be more than a few sackings over the course of the reformation.

Football has arrived at a fork in the road – and, as Yogi Berra, the American baseball player, famously said: 'When you get to a fork in the road, take it.' That is what football is doing – in small steps, some venturing this way, and others that – and all of them trying to come to grips with the advent of computers, data, analysts, research and more numbers than they can digest.

What may not be clear from *Moneyball* – the film that charts the rise of Beane's team – is just how dismal the Oakland A's were: an abysmal outfit playing in a drab stadium in front of a sparse, disinterested crowd. It was only in that environment that Billy Beane could find the motivation and, crucially, the *latitude* to change the club dramatically. 'We had nothing to lose,' Beane says. 'We were in a position where we could try anything and no matter what happened we were probably not going to end up any worse.'[1]

Often only desperation, marginality and a lack of money can create the conditions that support innovation. It is easy to draw the parallel in football. There, too, we can expect necessity to become the mother of invention.

Forecast 2

The football analytics movement will not feature a singular 'Bill James'.

Bill James and Charles Reep, for their personalities and their obsessions, were the perfect candidates to try to bring about fundamental shifts in the understanding of their chosen sports.

True innovations rarely spring from individuals inside the clubhouse. Outsiders can ask questions, they can query how things are done, and they see opportunities insiders don't. Because outsiders working in obscurity cannot be influenced very easily by those who don't know about them or those who don't believe in them, they can be the avant-garde – and a certain stubbornness and a huge amount of discipline and dedication will go a long way.

Where James and Reep differ is in their level of success: James was eventually hired by the Boston Red Sox in 2003, the year before they won the World Series for the first time since 1918. It was quite a leap for a man who, in 1977, had simply published a seemingly insignificant statistical pamphlet. By the time he joined the Sox, James's approach had been vindicated and imitated all around Major League Baseball, and in the next four years, the club would win two championships.

Reep, too, was brought inside the clubhouse, working as an

analyst for Brentford and Wolves, among others, but his transformation of the English game was associated with too few cups and championships to be considered a real success. That may be because of his own personal limitations, the quality of his data or deep differences between the sports of baseball and football. Whatever the true reason, Manchester City believe he fell short because of his limitations and those of his numbers: that is what led the club's match analysts, in the autumn of 2012 and with the support of Opta Sports, to take the unprecedented step of releasing an entire season's worth of match data to anyone with an email account who asked for it.

As Gavin Fleig, City's Head of Performance Analysis, explained to Simon Kuper: 'It's play by play, player by player, game by game, week in week out. I want our industry to find a Bill James. Bill James needs data, and whoever the Bill James of football is, he doesn't have the data because it costs money.'[2]

City's idea is well intentioned, but the idea of using crowds of analytics-minded fans to find 'Bill James' in the style of *The X Factor* may not work. The reasons are simple: *Moneyball* is already a Hollywood movie and everyone knows about it; more importantly, there is no analytical wilderness to explore. The Football Accountant has already left his footprints there. There is no undiscovered country in football's numbers as there was when James 'solved' baseball by inventing 'Runs Created' and 'Win Shares'.

Reep's failure also proves that the game is too fluid, contingent and dynamic for there to be a single winning formula discovered by a single great football mind. Instead, the multiple smaller insights generated by many will move the game forward.

Forecast 3

The volume of football data will increase by at least thirty-two times.

Reep had a lot more to tackle than James. At least the American had more than a century's worth of baseball box scores to work with, published in all American newspapers. Reep had to create the raw data of a match, and once he had done so, he had to store it on rolls of wallpaper.

As we have seen, Opta and StatDNA have partially computerized the gathering of football's numbers by employing analysts to code events from digital video of matches, while Prozone let cameras do the job for them. Together they have eliminated all traces of notebooks and paper from the storage of the numbers; today they are all in digital files. There is no reason to suppose that the leap from Reep's method to Opta's or Prozone's won't be matched by a similar leap in the coming decade.

It is a characteristic of the modern world of Big Data that if a number can be gathered, it will be, as cheaply and with as little human intervention as possible. The result of this hoarding is an explosion in the amount of data. According to *The New York Times*,

> There is a lot more data, all the time, growing at 50 per cent a year, or more than doubling every two years, estimates IDC, a technology research firm. It's not just more streams of data, but entirely new ones. There are now countless digital sensors worldwide in industrial equipment, automobiles, electrical meters and shipping crates . . . Data is not only becoming more available but also more understandable to computers. Most of

> the Big Data surge is data in the wild – unruly stuff like words, images and video on the Web and those streams of sensor data.[3]

Few things are more unruly than a football match with twenty-two players and a ball in almost constant motion (except for those Stoke games). There are two different ways in which more of the pitch's numbers could be gathered more cheaply. First, those same sensors we just mentioned. It's only a matter of time before players' kits and the ball itself will be fitted with GPS chips. The technology exists, and clubs are already experimenting with it in training. It may not come in the Premier League or the Bundesliga first; Brazil's Serie A or the US's MLS may be the ones to start. In fact, MLS has already partnered with sports equipment manufacturer Adidas to collect physical data through chips implanted in players' boots. This will produce a huge stream of positional data, and the chip in the ball will eventually obviate the need for goal-line video technology.

Some governing bodies will probably resist such advances, but the second means of gathering these data may make their resistance futile – crowd sourcing. Imagine, instead of Prozone's installed and fixed set of cameras which are very expensive, a handful of spectators scattered about the stadium with cameras embedded in their hats, scarves, or coats. As they watch the match, they record the action, and these video streams are later integrated and decoded through advanced software. Right now computers have a hard time distinguishing one player from another when they cross paths in a video, but soon they will have no more difficulty than the spectators.

As the cost of collecting football data plummets, more players in more matches in more leagues in more countries will be tracked. It's possible that the heir to our evergreen friend Jimmy Davies of Waterloo Dock may even begin to have access to computerized match reports. The surge in football's numbers will become a tidal wave, multiplying and doubling over this decade at least as fast as data in the world.

Forecast 4

Geometry – space, vectors, triangles and dynamic lattices – will be the focus of many analytical advances.

With the growing availability of positioning data and *x–y* coordinates for players and ball, analysts will be able to employ the mathematical tools of algebraic geometry and network theory to gain more insights into the game. The focus will move away from the ball and the counting of the 'ball events' that Reep first started collecting in his notebooks. Statistical attention will shift to players away from the ball, the clusters they form, the spaces they enclose and the way the ball and information move about the network they compose.

For this new stage of the reformation, Dynamo Kiev's Valeriy Lobanovskyi, with his interests in systems and space, will be recognized as a numbers game pioneer. Inspired by seeing the manager's bronze statue outside his eponymous stadium in Kiev during Euro 2012, Barney Ronay writes: 'He addressed football management as a wide-ranging empirical study, seeking informed scientific deduction about the more nebulous folk-football wisdom of his dugout contemporaries. For [him] the distillation of eleven competing blobs on a pulsing pre-modern computer grid appears to have also contained thrilling human variables, an applied chemistry to be grasped by study and fine adjustment.'[4]

The Ukrainian moved the grid from the computer to the pitch. He rigorously trained his players to perform like kinged pieces in draughts that could and would hop in any direction from square to square based on the movements of the other pieces. Although his club's play could sometimes be mechanical, it could also be incredibly effective as uncovered spaces, weak links and mistakes were minimized, and the imbalances of the opponent's defence were exploited. Positioning data will enable these 'draughty' practices and tactics to become more chess-like – sophisticated, creative and improvised.

When we look across the various sports, we can make a conjecture about the essential geometrical figure for analysis. Baseball is a game of ten points, a batter, a pitcher, a catcher and seven fielders, that are largely static, consistently positioned and rarely connected. Basketball, with its emphasis on the pick-and-roll, feeding a tall post player and the simple give-and-go, is a single, stretching and shrinking line segment connecting the two focal offensive players at any time during the game.

Football, as a more complex team game without a form of real possession, is largely about triangles. One such triangle might be the player currently touching the ball, the one about to receive it, and the off-ball player currently causing the greatest deformation in the defence's shape. Triangles might replace 'ball events' as the key unit of football analysis.

The use of networks to construct interacting webs of players and formations is already infiltrating the sport. Some of these efforts have already begun to identify the players that are central or peripheral to the passing network of a team on the pitch.

One recent study using Premier League data for two seasons beginning in 2006 found that teams whose passing networks were more centralized in one or two players scored fewer goals, even though these central players may have been the strongest players in the club.[5] Once again, balance, this time in the complex passing network, is a quality leading to success.

Forecast 5

There will be about 1,000 goals scored in the Premiership in both 2014 and 2024.

The power of numbers and of the models you can use lies in having many data points. Once we go from the specific – a pass, a match, or a player – to a bigger group – all passes, all matches, or all players – we can see patterns that are hidden when our nose is too close to the action.

There are two key issues the numbers game has to contend with: that football is defined by chance and by rarity. Chance, luck, randomness – not skill – accounts for much and possibly most of what happens in football. And the rarity of goals contributes hugely to that.

Goals have become rarer until, in recent years, they have levelled off at the same time as the game has become more equal on the pitch. Despite complaints about football being taken over by the super-rich, the data show us that football's long-term trend is towards ever fiercer competition. In the modern era everyone's more alike.

And at the very top, football production is the same. The best players seem to be interchangeable across leagues. This matters for how and where clubs recruit talent. If the very best

English players play football the same way the best Argentines do, then recruiting in Buenos Aires may sound glamorous but may not be particularly cost-effective compared to recruiting in Bristol, Leicester, or Preston. The sport in England and across the globe has settled into a competitive equilibrium.

Goals are wonderfully rare, but they will not get any rarer.

Forecast 6

The gap between the salaries and transfer fees of strikers and defenders and goalkeepers will shrink significantly.

Goals may not be the most reliable performance metric. When a team can do all the right things and can still end up losing, goals are not the best gauge of whether they played well.

Once you realize the power of chance and the value of a single goal for your team's fortunes, a few conclusions tumble out fairly easily. While football has always been enamoured with goals that are scored and victories celebrated, it has paid much less attention to goals that weren't conceded or losses that were averted. There are powerful human tendencies that explain this, but for budding performance analysts it means that, to fully grasp the nature of the game, we need to value defence properly. It is as important – and on occasion more important – than attack.

Understanding this has one very simple implication, one that will only be reinforced by the flood of new numbers: the salaries and transfer fees for defenders and strikers will become less skewed. As of today, the opposite trend still appears to hold. Using transfer costs from Paul Tomkins, Graeme Riley and Gary Fulcher's *Pay As You Play: The True Price of Success in the Premier*

League Era to calculate the relative value of different positions over the 1992/93 to 2009/10 period, the data show goalkeepers to be the cheapest position, with prices rising in increments as we move further up the pitch. Importantly, there does not appear to have been any narrowing of the gap in the cost of goalkeepers and defenders on one hand and strikers on the other. The ratio of striker to defender prices was 1.5 in the five seasons from 1992/93 to 1996/97, but 1.65 in the period 2005/06 to 2009/10.[6]

The hidden value of defence goes with a couple of other numerical insights. In football, more isn't always better – shooting more or having more of the ball is not an unfailing recipe for success. Sometimes less is more – less tackling, fewer corners and not conceding come to mind – while scoring the fifth goal counts for less than scoring the second. To put it in mathematical terms, football isn't linear or additive; it's multiplicative, dynamic.

Perhaps the most obvious manifestation of this is the way teams work together to produce wins, draws and losses. In basketball one superstar is 20 per cent of the starting team; football's giants make up a much more meagre 9.1 per cent of the total team. This means that the door is wide open for the team's poorest players and its most tenuous on-field relationships – the weakest links – to play a decisive role in determining a team's fortunes.

When *galácticos* and galoots have to cooperate to play winning football, it quickly becomes clear that there isn't one best way to play the game. More effective football comes in different flavours. And given the amounts of luck involved, you can play more successful football by following two broad strategies: either being more efficient, or being more innovative than your opponents. Both efficiency and innovation should bring more attention to the darker, defensive side of the pitch.

Forecast 7

The 4–4–2 will be replaced by the 150–4–4–2; organization will be the new tactics.

As Johan Cruyff knew, playing football with your brain allows you to outwit your opponent by thinking ahead a step or two or three. Football has always evolved with the times – albeit in fits and starts – and ultimately there has never been a way of halting its progression. The ball remains round. The game is played in more far-flung parts of the world by more people than ever before, by both sexes and on better pitches, with better equipment, by professionals who train to maximize their performance with the help of the latest knowledge in medicine, nutrition and computer science. As fans, we don't always see these changes at work – we don't watch players train, we know little about what they eat, or how they are monitored with modern technology. All we tend to see is what happens when the whistle blows.

But it's clear football analytics is continuing to infiltrate the game and change how managers, players, fans and executives think. So it's never been a question of whether analytics will be coming, but how clubs can best adapt to win.

The use of analytics has relatively little to do with specific statistics on players or teams. Analytics does not equal statistics; playing the numbers game is not really about numbers first and foremost. Instead it's a mindset and an information game – how much and what kind of football information clubs have; how they look at it, interpret it; and ultimately, what they do with it. There is no single truth; but there is an advantage you can gain by being smart about the information you use.

As the sourcing of players has become a global pursuit, with scouts and coaches scouring all corners for undervalued talent, the raw input clubs work with has become more similar. This convergence of the raw material of players at the top of the game and the worldwide diffusion of training and playing practices means that what will differentiate clubs will be their organization: which clubs and teams can organize themselves most effectively and figure out ways to try the new, unexpected and untested path to success. The history of innovation in football has been the history of tactics: better ways of organizing players in space and countering the opposition. But tactics is, at the core, about organizing your whole team – on and off the pitch – to maximum effectiveness.

When clubs are less differentiated with regard to their input – the quality of players recruited globally – football will become more of a team activity than simply the starting XI on the field. Instead all parts that help produce winning football on the pitch – the 150 or so coaches, nutritionists, physiologists, match analysts, scouts, you name it – will come into play more than ever before. And those clubs that function as teams of capable and willing learners, that can adapt to excel from week to week or minute to minute, will win out.

Flexible adaptation is the name of the game, be it with regard to the introduction of new technologies or countering an opponent's game plan on the fly.

Forecast 8

The current crop of absolutist managers are a dying breed. Once Arsène Wenger retires, all large clubs will have a General Manager/ Sporting Director – if not in name, then in function.

The person who in modern football has become responsible for overseeing a club's football side, deciding how to tread the path between efficiency and innovation, coaching its superstars and weak links and fighting football's unforgiving odds is the manager. He does matter, if not nearly as much as he himself seems to assume.

This is important: understanding the numbers means less need to rely on having played the game to become an expert in it. The power of numbers lies in the insight they can produce, but it is also a potent weapon in football's political combat. Information is power. It can bestow influence and take it away. Numbers and information mean transparency and meritocracy and so help do away with convention. The good news for fans is that the potency of randomness and chance in football will guarantee that the game will not change fundamentally – there will be plenty of room for underdogs to beat favourites, for riveting drama, and for the ball to hit the crossbar, fly skyward and eventually land in goal.

Once we realize that the defence and the weak links matter, the numbers will empower full backs at the expense of those well-paid forwards, weak links at the expense of stars and substitutes at the expense of the first XI. Once we can tell how a manager's choices really produce success on the pitch, the numbers game will empower clubs at the expense of managers.

Does all this mean the end of former players who become managers, to be replaced by geeks? Is this a moment of transition in football's long and glorious history, from the dictatorship of the manager to the unruly democracy of clubs run by vociferous fans and bench-warming malcontents?

We suspect not. Instead, it will become a more even partnership where managers are put in their place and become more cooperative team players in the club. Instead of being handed a transfer kitty to spend and facing accountability only when things go wrong, they will be forced to become part of the club's financial and organizational management. The new model of the modern manager will be Joe Maddon of baseball's Tampa Bay Rays.

What's needed is information and intelligence – managers that have both and know how to use them will succeed. There is every reason to believe that no manager of a top-flight club will be in sole command of football operations in ten years' time; rather, he will have an equal partner. This model is already popular in continental Europe, where the majority of clubs in Spain, Germany and Italy all employ Sporting Directors. Men like Monchi at Sevilla have forged reputations as masters of the market; some are former players – Matthias Sammer and Christian Nerlinger at Bayern Munich, or Marc Overmars at Ajax – while others have come from the backroom to work in recruitment.

Now that Ferguson is gone, once Wenger has retired, the age of the absolutists will be at an end.

Forecast 9

Just because a club does not play the numbers game will not preclude it from enjoying success; analytics will help you win, but so will money.

One might think that the flood of numbers alone and organizational changes will overwhelm all opposition to football analytics, but that would be to ignore the history of both innovation and revolution. Within ten years of the French

Revolution, Danton and Robespierre were both dead, and Napoleon was the head of state.

The number of technical staff has grown throughout the game and across leagues, while top-flight clubs all over the world have invested in match-analysis software coupled with video; some even use in-game software, like SportsCode. Clubs now have technical scouts like Hamburg's Steven Houston, men who scour the numbers before making a signing; they have sports and performance scientists like Manchester United's Tony Strudwick or Lille's Chris Carling; they have match analysts like Everton's Steve Brown.

Mobile devices and the internet have made the wall between clubs and the rest of the world more permeable. A community of bloggers has developed who use the internet to conduct their own analyses and some, like Onfooty.com's Sarah Rudd or Omar Chaudhuri from 5addedminutes.com have become professionally associated with the game – in Rudd's case with StatDNA and in Chaudhuri's case Prozone.

This is exactly what we should expect, Houston told us. His own experience working in basketball for the Houston Rockets of the NBA taught him that bloggers with analytical skills commonly ended up running the numbers for teams. As Bill James himself admitted in an interview with ESPN's Bill Simmons at the MIT conference, if the internet had existed in his time, he probably would have been a blogger.

These developments are exciting if you're into playing the numbers game from a sofa in your living room or in the club's front office. But this doesn't mean the picture is uniformly rosy. There are plenty of sceptics and pessimists within football, wishing these ideas would simply go away. As Real Madrid's Jorge Valdano explained in an interview with the German magazine *Der Spiegel*:

> You see, in my eyes, the pitch is a jungle. And what happens in that jungle has hardly changed over the past one hundred years. The thoughts that flash through a striker's mind today as he bears down on goal are the same ones that Maradona, Pelé and Di Stefano had in their day. What has changed is what surrounds the jungle. A revolution has taken place there, an industry has sprung up. We need to protect the jungle, to defend it from civilization and all of its rules. Civilization should be kept out of the game: Keep off the grass![7]

Whether this scepticism is genuinely felt or used as a smoke-screen to defend the established order, the obstacles to change are considerable. This resistance combined with the ever-present effects of chance and deep pockets mean that analysis will not be necessary for clubs to enjoy success over the next decade.

Forecast 10

The reformation of the counters will in turn be countered.

The old guard never goes without a fight, as Dean Oliver, basketball's numbers guru, knows. The author of *Basketball on Paper* and the first full-time statistical analyst in the NBA, Oliver worked for the Seattle Supersonics and the Denver Nuggets, both teams in a sport and a country where fans are familiar with and hungry for statistics. And even he told us that it was hard for analytics to find a true, permanent and accepted place inside clubs. Short-term pressures are too great and egos too big; those already in situ are too territorial. So Oliver left. Today, he is ESPN's Director of Production Analytics – shaping fans' knowledge and thinking about numbers across sport.

Football hasn't yet had its Moneyball moment, and whether it will is still an open question – basketball or American football or ice hockey haven't either. Whatever barriers Oliver encountered in American basketball are nothing compared to the walls that loom in football. Tradition is a potent impediment to anyone trying to introduce new ideas to clubs, trying to encourage their employers to play the numbers game.

As StatDNA's CEO Jaeson Rosenfeld explained to us: 'There is a system in place, existing power structures, ways that things have been done that need to adapt. That doesn't happen overnight. There are a lot of barriers; they have seen what happened at Liverpool, so they say, hey, Moneyball doesn't work in soccer. Humanity has figured out how to analyse more complex things than football. You never have an immediate success case. The things we're analysing now, it'll take a long time to figure out if we're right, and when someone does it, it'll take time to see if it's right. Once there's a bona fide success, they'll rush in.'

To use management language, this reluctance to tread where no one else has leaves football clubs analytically impaired. When asked to pinpoint the obstacles to growing the demand for analytics inside clubs, Rosenfeld identified two: 'I would say that the barriers – the scouts and managers – don't want to cede any of their authority.' Mark Brunkhart, President of Match Analysis, agrees. 'If you survey soccer coaches, you'll get the nod, "Yes, we believe in analytics work, we believe in the study of the sport" . . . Just because you have stats available doesn't mean anyone actually uses them to do anything.'

In baseball, the amount of data and the volume of insights into the workings of the game grew in a synchronized fashion. This is not true in football, which has seen its database move from a handful of bits to billions of bytes in a few short years, while breakthroughs remain few and far between. Curiously the twin incentives of companies to sell and of clubs to be seen as doing the latest thing conspire to produce mountains of data, but seldom generate real insight.

'All of us can sit in front of a pile of data and not learn anything from it,' says Paul Barber, the Brighton Chief Executive. According to Brunkhart, 'Because of *Moneyball*, there's this desire where people want to solve soccer. "Here, we're going to plug some numbers into an equation, it'll tell us what's wrong and we're going to fix things." If one more person comes to me and says, "We want to solve soccer, we're hiring an intern, can we have your data?" I'll cry. It's a very difficult thing to study.'

The unnerving prospect of a huge pile of numbers makes doing nothing a very attractive option for many managers and owners. They don't even know where to begin. Moreover, plenty of owners start behaving irrationally the moment they get anywhere near the pitch. As Keith Harris of Seymour Pierce told us, when it comes to making decisions based on evidence rather than gut, 'too many owners start taking off their business suits and putting on their track suits'. Finally, unlike American baseball teams – in a league with guaranteed membership and stable revenues – football clubs face a purer capitalist system: relegation and the shadow of bankruptcy and administration. This kind of downside risk makes most decision-makers more conservative and less likely to try new ideas. When jobs are at stake, traditionalists have an easier time digging in.

Opponents have a much easier time making their case that the numbers game doesn't work when someone like Liverpool's Sporting Director Damien Comolli openly talked about playing Moneyball on behalf of the club's American owners, and proceeded to splash money transfers for untested players who had relatively little impact on the club's fortunes.[8] Eventually, Comolli saw himself out of a job, the players relegated to smaller roles, sold, or sent out on loan.

Life After the Reformation

Today, well over two centuries after 1789, France tries to avoid mention of Robespierre as much as they try to avoid mention of the 2010 World Cup, or Ligue 1's status as the Premier League's unofficial feeder division. Prediction is notoriously difficult, especially about the future, to paraphrase Niels Bohr. Whether the numbers game goes the way of Robespierre – and is soon condemned and rejected – is anyone's guess. Perhaps it will do a Napoleon, and burn briefly, but brightly. Maybe it will do neither, instead evolving more gradually.

Yet we are certain about what will happen in the long run. Just as France was destined to become a democracy, in football innovation and technology will win out. The best managers, players and clubs will adapt and win; football analytics will play a critical part in the game.

If we look closely, we see the future of analytics happening in our living rooms. Chris has two sons, aged ten and thirteen. Like most boys their age, they spend a substantial portion of their free time playing on football video games. They argue over whether one of them should buy or sell a certain player for his team and for how much, based on their performance stats and their potential to make an impact for their club.

Chris's older son, when asked by his maths teacher to demonstrate with an example from the real world the use of percentages and trends in data, chose the scoring rates of

Lionel Messi and Cristiano Ronaldo. His friends had done the same, choosing Neymar and other players.

These children will grow up to watch football, play it, love it and perhaps even manage it. They like numbers; they think about them; they know them. They are believers in the game, and consumers of data. They will marvel that there was ever a time when football was free of numbers, resistant to analysis, and reluctant to reform.

Extra Time – The Numbers Game at the World Cup

A glance at the history books makes you think the World Cup, put simply, defies logic.

Football's great jamboree of north, south, east and west brings together small and large, the fearful and the fabulous, the happy-go-lucky and the win-at-all-costs. What ensues is chaos. The story of the World Cup is a story of upsets and underdogs, of the Miracle of Bern and the Maracanazo. It is the stage on which the United States' motley collection of enthusiastic amateurs toppled England in 1950, and North Korea obliterated Italy in 1966. It is the place where Algeria can beat West Germany and Cameroon overcome Argentina. The World Cup is the ultimate level playing field. This is where every dog has its day. This is where miracles happen.

If one of the World Cup's key characteristics is its unpredictability, is there a place for the numbers game? Is there a pattern in the frenzy? We have already used data garnered from the tournaments, of course: studies using the World Cup as their sample have helped us see that winning teams were more reliant on passing the ball than losing sides, and that more managers of World Cup sides timed their substitutions best. That is not all, though. Far from it.

The story of the World Cup is full of romance. That's why it captivates the planet. But it turns out that there is a signal in the chaos.

Ensuring Victory: The Quality of Ruthlessness

Héctor Chumpitaz remembers wondering if the two men had opened the wrong door.[1] He and his teammates in the Peruvian national side were busily going through their final preparations for the match: dressing, taping, worrying, praying. The last thing they expected to see was General Jorge Videla, President of Argentina, and the former United States Secretary of State Henry Kissinger. They said they were there to wish the players good luck, and let them know that the Argentine public was expecting a fine game. It was a peculiar message. Peru, after all, were facing Argentina that night. They were all that stood between the hosts of the 1978 World Cup and a place in the final.

That was by no means guaranteed. That tournament had seen the remaining eight nations sent not into traditional quarter-finals, but two final groups of four. Argentina and Peru were in with Brazil and Poland. Going into that final game – no simultaneous kick-offs back then – Brazil were top, with two wins and a draw, and a goal difference of plus five. Argentina needed to beat Peru by more than three goals to overtake their bitter rivals and secure their spot against Holland in the final.

By half-time, they were almost there. The hosts were two goals up, due in part to the fact that Peru had missed what Sir Walter Winterbottom, the former England manager, called 'the best chance of goal-scoring I had ever seen'.[2] They needed the same again to qualify. Peru, thankfully, were in an obliging mood. Five minutes into the second half, Argentina were four up, against Chumpitaz – considered one of the best defenders in South America – and a midfield rated by the respected magazine *El Gráfico* as 'the best in the world'.[3] More bizarrely still, Peru immediately substituted one of their key players, José Velásquez, for no apparent reason. Argentina ran out 6–0 winners.

That game is the focus of one of the great World Cup conspiracy theories. Most assume it was a fix, though nobody has solid proof. There are suggestions it was purchased with 35,000 tons of grain and a guarantee to unfreeze $50 million worth of credits. Others believe Videla and Argentina's brutal military junta had threatened to torture and 'disappear' Peru's players. Some think it was more straightforward: a straight cash bribe. Or maybe it was the dementor-like presence of Videla and Kissinger which unsettled Peru. Whatever happened, whatever is true, it worked: Argentina beat the Dutch, minus the absent Johan Cruyff, to win their first World Cup.

It was not the score, though, that alerted observers to the fact that there was something fishy going on. It was the fact that there was no great gulf in quality between the sides. Seeing a team win 6–0 in the World Cup group stages is hardly unusual. History is littered with occasions when the big have administered hearty beatings to the competition's lesser lights. Hungary have made a habit of running up cricket scores in the early rounds: from their 6–0 win against the powerhouse of the Dutch East Indies in 1938, through the 9–0 humbling of South Korea in 1954 to the 10–1 win against El Salvador in 1982, the Magyars have rarely been short of goals.

The intriguing thing is that on none of those occasions did Hungary go on and win the tournament – though they went as close as possible in 1954. In the major European leagues, goal difference is closely correlated with not just the ability of a team, but its eventual success. The anecdotal history suggests that does not automatically follow in the World Cup. Scoring a lot of goals in the group stages does not necessarily mean you will go on to win the thing – unless, perhaps, there is the shadowy hand of a dictatorship at play.

There are several explanations for why this should be. There is the fact, first of all, that the World Cup is a relatively small sample in statistical terms. Between 1930 and 2010, there have been 773 matches in total, a number that is equivalent to two seasons in the Premier League, La Liga or Serie A.

Moreover, the World Cup does not work like a league, where the same teams and players – more or less – play in back-to-back seasons. Teams and players appear and disappear from one World Cup to the next. So, too, does structure. Goal difference has long been a reliable measure of team quality in league football because of the consistency in the structure of the seasons – the same number of clubs play each other home and away for years on end. This makes analysis easier. By contrast, the format of the World Cup has only settled down – with thirty-two teams, eight groups of four, and the knock-out stages beginning with the last sixteen – since 1998. On a number of levels, the World Cup provides not just a small sample, but a noisy one. Reliable trends are far harder to discern.

There is one other problem inherent in the group structure when you attempt to use the early games of a tournament to establish how good a side actually might be. It was not true of Argentina and Peru in 1978, but often the third and final matches in a group might not be as important as the first and second. A team that has won its first two games might have nothing riding on the third. They might make wholesale changes. They might not try quite as hard with the pressure off.

The conclusion is obvious. It is a small, noisy sample, full of dubious results, dead rubbers and red herrings. The early games of the World Cup are no way of determining who has the best chance of success in the tournament. Who does best in the group stage is not a reliable indicator of who will win in the end. Is it?

The numbers say different. Just as it is in league football, goal difference is a hugely important statistic in the world's greatest football tournament.

There have been sixteen World Cup finals since the competition resumed in 1950. In eight of those games, one finalist had a larger goal difference going into the game. On seven of those eight occasions, the finalist with the better goal difference won. The only exception is 1974, when Cruyff's Dutch side overwhelmed West Germany, but somehow lost 2–1.

There is more. It is possible to look at how well a nation's goal difference in the first round predicts whether it will reach the semi-finals, once the knock-out rounds have begun. A total of 180 national teams have qualified for the second round since 1954, with an average margin of victory just a shade below a goal per match. Using a statistical method that relates performance and probability, we find that a larger margin of victory in the first round is significantly related to a better chance of making the semi-finals. In fact, an extra goal in each group-stage game moves the average nation's probability of making the semi-finals from 25 per cent to 40 per cent.

Even those games between a nation which has already qualified from the group and one which will eventually be eliminated are telling. As Figure 54 shows, the more goals you score against a minnow as you send them on the way home, the more likely you are to make the semi-finals.[4]

Goal difference, then, is not just a measure of quality in the club game. The same goes for national teams. If you are trying to predict the winner of the World Cup, it may make sense to wait until the group stage is over. The team bound for glory will be the one that has least mercy on the teams who are just happy to be there.

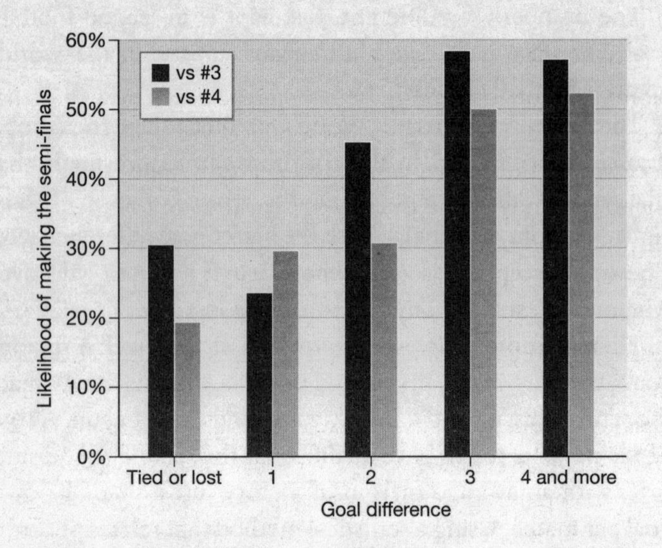

Figure 54 Beating the weak first-round teams and qualifying for the semi-finals

Calculating on the Spot

There should be no part of football that is easier to analyse than penalties. After all, even the most ardent detractor of analytics would find themselves on shaky ground suggesting that a penalty shootout is 'too fluid' to be subjected to the sort of analysis that has been so successful in baseball. The situation is exactly the same: like a baseball pitch, a penalty revolves around a confrontation between two players who can take just one action: the pitcher pitches, the taker shoots, the batter swings (sometimes) and the goalkeeper dives. Penalties should be simple to investigate. No wonder they have proved attractive to a variety of researchers looking to establish what makes a penalty – or penalty save – successful.

This is good news, because penalty shootouts are getting more and more common. We have already seen how there has been a dramatic convergence in top-level football as best practice within the game has diffused around the globe. This phenomenon is borne out at the World Cup: the greater the diffusion, the more competitive matches; the more competitive matches, the more draws.

Before 1978, any knock-out games in the World Cup that remained level after ninety minutes of normal time and thirty minutes of extra time would be replayed. After 1978, these matches would be resolved by penalties. There were forty-two knock-out games between 1954 and 1974. There was not a single replay. There have been 118 knock-out games since 1978. There have been twenty-two shootouts. That almost 20 per cent of these games are level after 120 minutes reflects that football has converged.

More teams play the game in the most effective manner. The margins between teams are growing ever more slender. Penalties are becoming more and more important. It is often the moment when heroes are made. Increasingly, one of those heroes is an analyst.

Back in 1970, penalties were introduced in a bid to reduce the role of chance in determining who went through in knock-out ties. Before then the replay had not been the only way of settling drawn games. There was another option: the drawing of lots.

It seems remarkable now, but very occasionally that is how games had to be settled. In 1954, that is precisely what happened to Turkey and Spain. A second replay had ended level, so the fourteen-year-old son of a groundskeeper was blindfolded and asked to pick a ball from a small urn.[5] Spain had gone into the game as heavy favourites; now it was down to a 50/50 shot. They were convinced the fates were against them: '[we thought] that there was nothing more we could do, that everything had gone so badly that the kid wouldn't pick our name.'[6] They were right. Turkey went through.

We often talk of penalties being a lottery – we will come to that later – but at least there is some degree of skill involved. It is not naked chance. That does not mean, though, that it is perfect.

Take the infamous quarter-final between Ghana and Uruguay at the 2010 World Cup. The game is in extra time, the scores level at 1–1. We are in the last minute of added time. Ghana win a free kick on the right. It sails into the box. It is headed back across goal. Stephen Appiah volleys it towards the net. Luis Suárez is standing on the goal line. The ball hits his knee. The ricochet takes it on to Dominic Adiyiah's head. His effort is on target. Pause the game there. At that point, Ghana's chance of qualification is 100 per cent.

Suárez sticks out both hands, shoving the ball away from goal. Olegário Benquerença, the referee, has no choice. He reaches into his pocket for the red card. Pause the game there. At that point, Ghana's chance of qualification has been reduced to 75 per cent – the average conversion rate for a penalty kick. Suárez – widely condemned as a cheat and a villain – had not only done what all professionals would do in that circumstance, he had done what he should have done. He gave Uruguay a glimmer of hope. Asamoah Gyan missed the spot-kick, smashing it off the crossbar. Penalties. Benquerença takes a coin from his pocket and tosses it. It goes in Uruguay's favour.

They choose to shoot first. Pause the game there. Ghana's chance of qualification now stands at just 40 per cent. In the space of a couple of minutes, no more, the African side have gone from shoo-ins to become the continent's first ever representatives in a semi-final to considerable underdogs.

It is thanks to one of those heroic analysts that we know this. It is a familiar face, too: much of the best work in the area of penalties has been carried out by Ignacio Palacios-Huerta, the same professor who diligently counted the numbers of goals scored since the early decades of association football. Palacios-Huerta has done much the same for spot-kicks, tabulating more than 10,000 penalties between 1970 and 2013. He has identified the key moment in any shootout. It is the coin toss.[7]

The idea that penalties removed the role of chance in deciding knock-out ties is an illusion. Its role is reduced, but a substantial amount still rides on that one moment of sheer, naked luck. Out of the thousand shootouts in Palacios-Huerta's database, the team that gets to shoot first has won just over 60 per cent of the contests.[8]

Choosing to shoot first is not the only piece of advice teams can glean from the research into penalties. This will be of particular concern for England, whose record in shootouts remains a source of a remarkable national inferiority complex. Steven Gerrard's description of the pressure he felt before his spot-kick against Portugal in the 2006 World Cup speaks volumes for how England's chronic history of failure from twelve yards plays on the country's minds.

> Now the spotlight burned on me. Here goes. I broke away from the safety of my friends in the centre-circle. Suddenly I was alone, making my way towards the penalty spot, towards my

> fate. The journey was only forty yards, but it felt like forty miles . . . As I neared the spot, my body went numb . . . I went through my penalty routine. Set the ball right? Yes, done. Remember all the good kicks in training? Yes . . . Know where you're going to place it? Yes. Ricardo's good, but if I place the ball exactly where I want, at the spot where Robbo, David James and Scott Carson told me about in training, Portugal's keeper can't stop it.[9]

Gerrard, sadly, did not place the ball exactly where he wanted. Gerrard missed. So did Frank Lampard and Jamie Carragher. England did not even take the shootout to the fifth man. The national shame deepened a little more that night in Gelsenkirchen.

Watching footage of those painful few minutes, many of the researchers into this area would not have been surprised by England's travails. There has been extensive work done on the body language of penalty-takers, and the haste with which they take their kicks. Lampard and Carragher both turned their backs on Ricardo instead of backpedalling from the spot after placing the ball. Carragher was in such a panicked rush he took his kick too quickly, and was forced to have another go. Those analysts would tell you that these were critical clues in giving Ricardo an advantage. They would tell you that the goalkeeper believed Lampard and Carragher were likely to miss, so he waited a beat longer before diving, giving him a slightly better chance of blocking each shot.[10] England's takers had denied themselves slight advantages.

There are a number of small tricks for the goalkeeper, too, in a bid to make sure they can join Ricardo, Cláudio Taffarel and the others in the World Cup's litany of heroic shot-stoppers. First, it has been shown that his very presence is distressing to the penalty-taker. One experiment showed that a keeper could stand six to ten centimetres off-centre in the goal and induce 10 per cent more kicks into the wider area, meaning his dives in that direction are more effective.[11] He can also make the kicker believe he is bigger than he is by holding his arms out and above his shoulders.[12] This provides a human equivalent of the famous Müller-Lyer illusion, which makes you think that the vertical line on the left in the diagram is longer than the one on the right. And then there is the technique that both Bruce Grobbelaar and Jerzy Dudek would recognize: he can dance. Waving your arms or generally clowning around can attract the attention of the kicker away from both the target and his eventual effort.[13]

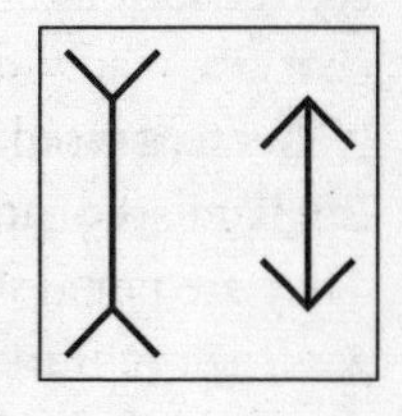

The background to these tricks is the goalkeeper's very existence. We have already introduced you to the idea of the Maldini Principle: our brains prioritize something that is present over something that is absent. A goalkeeper in a penalty shootout is very much present. The target area Gerrard described is absent. The former is real, the latter imaginary. The goalkeeper naturally draws the gaze of the taker – a phenomenon borne out by studies[14] and exacerbated by anxiety, time pressure and a direct instruction not to look at the goalkeeper, whatever he is doing with his arms.[15]

Of course, in the case of Gerrard, none of that should have been relevant. The Liverpool captain did exactly what he would be recommended to do: he walked backwards from the spot to take his kick. He did not show any haste. He had a 'goalkeeper-independent' strategy that involved him shooting to a specific spot, whatever Ricardo did. And yet he still missed. This is the most depressing conclusion for England. The reason an accomplished penalty-taker like Gerrard erred in Gelsenkirchen may well have been the shirt he was wearing.

The Norwegian sports psychologist Geir Jordet and his colleagues have examined whether a team's recent record in a major shootout predicts whether they will win or lose their next one. Jordet finds that national teams that lost their previous shootout converted fewer of their kicks (66 per cent) compared to teams that were victorious previously (85 per cent conversion) and teams that hadn't been in a shootout (76 per cent).[16]

Gerrard would be present for another shootout defeat with England six years later at the European Championships, too. They are not alone: Ghana, beaten on penalties by Uruguay in 2010, lost to Burkina Faso in the 2013 African Cup of Nations in the same fashion. Their records suggest that they will only convert three of five kicks the next time they are involved in a shootout. The numbers make it clear to their managers that selecting the players who had missed previously would be a recipe for disaster. Fresh blood is needed, players unencumbered by their own memories of personal failure.

Roy Hodgson, Gerrard's manager with his national side, is aware of this. He planned before the World Cup to bring in the renowned sports psychologist Doctor Steve Peters to speak with his players to prepare them for the tournament. He knows how much pressure the players are under. Peters is intended to help them quieten their inner chimp: the part of the mind he believes is responsible for nervousness and anxiety. Maybe it will help.

If it comes to it, Hodgson should probably not repeat his words that accompanied Peters's hiring: 'One forgets sometimes how important these tournaments are and what big occasions they are. You don't get that many shots at it and you have a lot of time to regret if you don't give it your best shot.'[17] Rather, he should have had his coaches and analysts searching the numbers for small advantages and tendencies. But this being England, the numbers will have had one clear message for Hodgson: hope it doesn't go to penalties.

Acknowledgements

It was 1974. My best friend and I were eight, two boys captivated by the World Cup being played in West Germany, following every twist and turn of the tournament's progression with intense interest.

The football was just one side of it. We were equally fascinated by gathering the collectable cards published for the tournament, using them to compare Germany's Franz Beckenbauer to the Netherlands' Johan Cruyff, or Sepp Maier to Poland's awe-inspiring keeper Jan Tomaszewski.

But we had another, more analytical use for the cards: we wanted to know which players were most famous, and which ones most popular. So we took pencil and paper and made our way to the town square. There, overcoming our shyness, we approached passers-by and asked them a simple question as we showed them a random sample of pictures with players on them:

'Do you know who this is?'

If they said no, we let them go and marked the player's name on our sheet with a no. If they said yes, we asked them whether they liked the player, and we put a mark on our sheets next to his name if they said yes.

I don't remember who 'won', but I'm pretty sure it was Beckenbauer or Gerd Müller – the Bomber; certainly not Paul Breitner, too counterculture for this sleepy conservative town. That year was also the summer I started playing football, re-enacting the day's matches in a small alleyway near my house with friends. I would always end up playing in goal.

For Germans, 1974 was a memorable anniversary – twenty years after the Miracle of Bern, another World Cup when a West German side beat an allegedly superior team, with Holland in the role occupied in 1954 by Hungary.

A few years later I met Fritz Walter, captain and hero of West Germany's conquering 1954 side when he came to visit with a team of youngsters selected to represent the region's football association in Koblenz. I was overawed – even more than Konrad Adenauer, West Germany's first post-war Chancellor, Walter personified West Germany's becoming a new country. So 1974 was the year I became a football player and – in hindsight – the year I became an analyst, too. I ended up playing, as a goalkeeper, for a number of years, earning a few Deutschmarks along the way. When I realized I was not cut out to make it as a professional, I hung up my gloves and went to university instead, eventually earning a PhD and becoming a professor at an Ivy League university in the United States. For over twenty years, I taught political economy and political sociology. At that point, the numbers game was far from my mind.

When Dave and I started talking about football, it was just one of those chats about Stoke's Rory Delap and his spectacular throw-ins. Dave is a neighbour and friend, a fellow academic and economist. But more than that, Dave, growing up steeped in basketball and baseball, is another who found himself collecting cards as a child, drawn as much to the pictures of the stars on the front as the mountain of statistics on the back.

In his case, the cards in question were the 1969 edition of Topps' baseball series. Dave's beloved Chicago Cubs, that year, had a talented team with an infield of Ron Santo, Don Kessinger, Glenn Beckert, Ernie Banks and Randy Huntley, but withered as the season wore on and eventually lost the Eastern

Division to the New York Mets. Every dime Dave had went on packs of those cards in the hope of collecting the entire Cubs' roster; not simply to have pictures of his heroes, but for the statistical tables on the reverse side, just the kind of mound of data a maths-loving kid could pore over.

The obsession with numbers never disappeared. Later, as a southpaw pitcher for Harvard, Dave would study his own statistics before and after starts and he would chart the pitches of his teammates when they were on the mound. Not being blessed with an overpowering fastball, he had to be crafty and analytical. The numbers gave him an edge.

It should be no surprise that, to a former pitcher, Delap's powerful hurling of the ball was eye-catching; to the analyst in him, the natural reaction to such a phenomenon was to prompt and probe and question.

The trouble was, I didn't have too many good answers. So off we went, talking more frequently and then more seriously – about football, and about football numbers. Why teams win and lose; how you can spot a good player or manager; what it is that makes football football. The end result is this book.

In the process of working together, we have figured out one simple truth: Just as it takes a team to win a football match, it takes a team of great friends and colleagues to write a book. A community of people helped us by being generous with their time, as well as honest, kind and helpful. In the process of writing this book we incurred a good number of debts that we happily acknowledge here. None of them share any of the blame for our mistakes, since it's become readily apparent to us that we are the weak links in this network of very smart people.

We are grateful to all those who graciously and generously gave of their time to help us understand various facets of the professional game, football analytics and the history of the numbers game, including Duncan Alexander, Peter Ayton, Rob Bateman, Matthew Benham, Amit Bhatia, Jonas Boldt, Nick Broad, Steve Brown, Mark Brunkhart, Andy Clarke, Phil Clarke, John Coulson, Gabriel Desjardins, Matt Drew, Michael Edwards, Gavin Fleig, Gary Fulcher, Simon Gleave, Ian Graham, Paul Graley, Howard Hamilton, Keith Harris, Steven Houston, Dan Jones, Don Kirkendall, Simon Kuper, Mitch Lasky, Keith Lyons, Scott McLachlan, John Murtough, Boris Notzon, David Paton, Kris Perquy, Richard Pollard, Clive Reeves, Graeme Riley, Jaeson Rosenfeld, Sarah Rudd, Robin Russell, Ishan Saksena, Barry Simmonds, Zach Slaton, James Smith, Rod Smith, Stefan Szymanski, Paul Tomkins and Blake Wooster. Thanks also to all those who wished to remain anonymous – you know who you are.

It would be impossible to write a book on football as the numbers game without access to data. We are grateful to the open-handed people at the leading data providers, who took a respite from the fierce competition to supply us with very high-quality numbers and much encouragement in our research and writing: Matt Drew and John Coulson at Opta, Jaeson Rosenfeld at StatDNA and Simon Gleave at Infostrada. Thanks also to Tony Brown of soccerdata.com who provided the English football league data we used in Chapter 2.

A million thanks to our amazing agent, David Luxton. We are deeply indebted to David's unerring judgment and diplomatic touch – thanks for taking a gamble on us and for repeatedly saving us from ourselves! Many thanks also to the midwives (Raphael Honigstein and Jonathan Wilson) who facilitated our collaboration with David. Their *Inverting the*

Pyramid and *Englischer Fussball* and other writing continue to be an inspiration!

We are deeply indebted to editor extraordinaire and brilliant wordsmith Rory Smith. Without Rory, this book would have been a meandering, overly technical, academic journey through Chris and Dave's brains, rather than a football book. He is one of football journalism's most impressive minds.

Although this title isn't a biography of an ex-player, Joel Rickett at Penguin wasn't afraid to think there might be potential in a book about football numbers. We obviously think he made the right call, possibly based purely on instinct rather than the numbers, and clearly a willingness to defy convention (along with a deep curiosity about corners). Thanks also to Joel's colleague Ben Brusey for giving the book such a careful read, and to Trevor Horwood for helping us keep our final mistake count below Lobanovskyi's threshold.

We are grateful to colleagues and friends who answered lots of seemingly random questions and provided plenty of helpful hints on short notice without compensation other than a heartfelt 'Thank you!' They are Tom Gilovich, Raphael Honigstein, Ben Sally, Bryce Corrigan, Robert Travers, Pete Nordsted, Simon Hix and Kirk Sigel.

For comments and feedback at an early stage, we thank David Rueda and Derek Chang. Without Derek this book wouldn't exist.

Many thanks also to the ever cheerful Ramzi Ben Said and Judith Ternes who provided excellent research assistance along the way. We are also grateful to Stephanie Mayo for helping us track down clubs' financial data.

We would like to thank the office away from the office, Ithaca Bakery at Triphammer Mall in Ithaca, NY, where many of the ideas in the book were concocted over bagels and coffee.

Chris would like to thank the staff at Gimme Coffee on Cayuga, the Amit Bhatia Olin Library Café, The Shop and The Big Red Barn for patience with this particular customer who spent days on end sitting and typing away in a corner, but consuming relatively little coffee. Dave would like to thank his dog and two cats for making room out on the porch so he could write.

Finally, we are grateful to our own teams: Kathleen O'Connor, Nick and Eli Anderson, Serena Yoon and Ben, Mike, Tom and Rachel Sally. Our home field advantage lies solely in their enthusiasm, motivation and unwavering support – and, truth be told, the occasional taunting chant, one that we know is always full of love.

Notes

Football for Sceptics – The Counter(s) Reformation

1. http://en.wikipedia.org/wiki/Analytics
2. Simon Kuper, 'A Football Revolution', *Financial Times*, 17 June 2011; www.ft.com/cms/s/2/9471db52-97bb-11e0-9c37-00144feab49a.html#axzz1qzPfmj6H
3. Matt Lawton, 'Roberto Martínez – The Man Who Shook Up the Season', *Daily Mail*, 20 April 2012.
4. *Globe and Mail* (Toronto), Friday 13 May 2011.
5. Cited in Lyons (1997).
6. Pollard and Reep (1997), p. 542.
7. The origins of academic statistical analysis of football data date back a bit further, to another Englishman, by the name of Michael Moroney, whose 1951 book *Facts from Figures* included an analysis of the numbers of goals scored in 480 games in English football to understand if they follow predictable patterns (Brillinger, 2010).
8. Simon Kuper, 'A Football Revolution', *Financial Times*, 17 June 2011; www.ft.com/cms/s/2/9471db52-97bb-11e0-9c37-00144feab49a.html#axzz1qzPfmj6H
9. See the *Journal of Sports Sciences*, October 2002, special issue on performance analysis, for an overview of how the field has evolved. For more insight into match performance analysis and match analysis, see also Reilly and Thomas (1976), Larsen (2001), McGarry and Franks (2003), and Hughes (2003).

10. Ayton and Braennberg (2008). Focusing on 1–1 ties eliminates those instances where a very weak team happens to score first and then is overwhelmed by a more powerful squad. The ideal experiment would be cloned squads of equal talent where there was a triggering goal by one side to see how the other side responded (that is, when the only difference between the squads is the fact of the goal itself).
11. Vialli and Marcotti (2006), p. 155.
12. On average, the teams in the sample took 5.4 corners per match, consistent with the long run average of 5.5, and the clubs earned between 4 and 6 corners in the average match. Shots and goals created from this particular match situation are defined here as occurring within three touches of a corner.
13. There also is considerable variation across clubs on this. At the low end, some of the very best clubs in the league managed relatively few shots – about 1 to 1.5 in 10 – relative to the number of corners they produced. In contrast, some of the worst teams in the league produced a relatively high number of shots in the aftermath of corners (Chelsea were the exception), at a rate of 1 in 4 or even 1 in 3 (West Ham and Stoke).

Chapter 1. Riding Your Luck

1. Press Association, 'World Cup Final: Johan Cruyff Hits Out at "Anti-football" Holland', *Guardian*, 12 July 2010; www.guardian.co.uk/football/2010/jul/12/world-cup-final-johan-cruyff-holland
2. Christian Spiller, 'Der Fußball-Unfall', *Die Zeit*, 20 May 2010; www.zeit.de/sport/2012–05/champions-league-finale-chelsea-bayern
3. http://m.guardian.co.uk/ms/p/gnm/op/sBJPm4Z87eCd_ev4Q2pP53Q/view.m?id=15&gid=football/blog/2012/may/20/roberto-di-matteo-roman-abramovich&cat=sport

4. www-gap.dcs.st-and.ac.uk/~history/Biographies/Bortkiewicz.html and http://statprob.com/encyclopedia/LadislausVonBORTKIEWICZ.html
5. Poisson's technique is still widely used by statisticians. In his book, *Das Gesetz der kleinen Zahlen* (The Law of Small Numbers) (1898), Bortkiewicz constructs a Poisson distribution for each corps and then sums over all the corps to get an even more accurate match between actual and estimate.
6. For the statistically minded: let λ be the base rate, then the probability that the number of events equals some number, *n*, is: $Pr_{\lambda}\{X = n\} = \frac{\lambda^n e^{-\lambda}}{n!}$. Furthermore, the events have to be mathematically rare, random and independent.
7. This doesn't mean that bookmakers have to have a good understanding of football – all they need to know is what makes the odds better or worse in any one match. But bookmakers also have to compete for their customers' business, so they have a strong incentive to get it close to right most of the time. Of course, the odds they offer are adjusted slightly in their favour from the estimated 'true' odds as they see them – otherwise their business wouldn't be profitable; nevertheless, the old 'two heads are better than one' maxim would lead us to believe that the combined odds set by numerous bookies can be taken to be an indicator of how predictable the bookmaking fraternity assume outcomes to be.
8. We collected the data from oddsportal.com
9. For simplicity, this discussion ignores bookmakers' profits.
10. We can do this by dividing 100 by the decimal odds. For example, decimal odds of 2.0 transform to a 50% chance of winning (100/2.0). So an NBA team with odds of 1.25 is judged to have an 80% (100/1.25) chance of winning.
11. In a similar analysis of the 2007/08 Bundesliga and Premier League seasons, Quitzau and Vöpel (2009) found that chance

played a role in 52.7% of all Bundesliga and 49.5% of all Premiership matches.

12. Ben-Naim, Vazquez and Redner (2006).
13. Heuer, Müller and Rubner (2010).
14. Technically, the idea is that goals might not be independent and that the base rate is altered by the number of prior 'events' in a match.
15. Skinner and Freeman (2009).
16. For the mathematical details, see http://understandinguncertainty.org/node/56
17. In Spiegelhalter's example, the variance of the actual league points at the end of the season was 239, compared with the theoretical variance of 61 had all the teams been of equal quality and the results of the matches essentially due to chance. Since 61/239 = 0.26, 26% of the variance in the Premier League points is due to chance. The standard deviation of the observed points, which is the square root of the variance, is roughly equal to 15, while that of the random points is around 8. This means that the observed points have about twice the range of the random points, so about half the spread of points is due to chance alone. For the details, also see http://understandinguncertainty.org/node/61
18. Oliver Fritsch, 'Bei zwei von fünf Toren ist Zufall im Spiel', *Die Zeit*, 22 December 2010; www.zeit.de/sport/2010-12/zufall-fussball-pokal-bayern
19. The conditions are these: the shot was deflected; the goal was scored on a rebound; the ball hit the post or crossbar before going in; the goalkeeper touched the ball and had a reasonable opportunity to make the save; the goal was scored from a far distance; or the ball was gifted to a striker in front of the goal by a bad pass.
20. Lames (2006).
21. www.sport.uni-augsburg.de/downloads/30_verschiedenes/Pressemitteilung Zufall.pdf

22. Tobias Hürter, 'Alles Zufall', *Süddeutsche Zeitung*, 8 June 2006.
23. www.sport.uni-augsburg.de/downloads/30_verschiedenes/Pressemitteilung Zufall.pdf
24. www.dailymail.co.uk/sport/football/article-1280360/Louis-van-Gaals-plan-silence-star-pupil-Jose-Mourinho.html
25. *The Blizzard*, Issue 1; theblizzard.co.uk

Chapter 2. The Goal: Football's Rare Beauty

1. Will Springer, 'A Day When Scottish Football Scorched the Record Books', *The Scotsman*, 9 December 2005; www.scotsman.com/news/arts/a_day_when_scottish_football_scorched_the_record_books_1_466092
2. Fraser Clyne, 'Arbroath Legends'; www.arbroathfc.co.uk/history/36-0-team.htm
3. In 2010 the Ballon d'Or and the Fifa World Player of the Year were merged into one award.
4. Hughes and Bartlett (2002); see also Read and Edwards (1992).
5. Palacios-Huerta (2004).
6. Some leagues started later than others, and there were interruptions because of two world wars, of course. For details, see Palacios-Huerta (2004), p. 244.
7. Colvin (2010), pp. 8, 9.
8. We are attempting to hold all other factors constant and focus just on the effect of greater skill on goal scoring and goal prevention. If scoring declined greatly in the first tier due to, for example, much more skilled goalkeepers, whose abilities are proportionally so much greater than their fourth-tier counterparts now than they were half a century ago, then goals – unsuccessful saves – should have declined much more in the top tier than the fourth tier.

9. Wilson (2009).
10. This is a quote attributed to Pierre Teilhard de Chardin, the father of the idea of convergent evolution.
11. Galeano (2003), p. 209.
12. See Miguel, Saiegh and Satyanath (2011). Aside from the exposure to civil war, we can of course imagine a number of alternative explanations. Perhaps they have a more strongly developed need to fight for a place in the starting XI, given the relative poverty of their youth and the dependence of others on their income. We also should think about the role of referees in all this. Assuming that referees are not immune to ethnic stereotypes, we can imagine that refs call systematically more fouls on players from certain regions or with certain visible characteristics. In the NBA, for example, more fouls are called on black players than on white ones. On the topic of referee stereotyping of players in football, see Gallo, Grund and Reade (2013).

Chapter 3. They Should Have Bought Darren Bent

1. Dilger and Geyer (2009). Two other papers came to similar conclusions: Garicano and Palacios-Huerta (2005) examine a couple of years of Spanish football data while Brocas and Carrillo (2004) build a game theory model.
2. Here are the values of points per goal. 0 goals: 0.28 points; 1 goal: 1.13 points; 2 goals: 2.12 points; 3 goals: 2.67 points; 4 goals: 2.90 points; 5 and more goals: 3 points.
3. An alternative would be to condition all these values on the actual match score at the time. However, it is not clear why this would be a preferable analytic strategy since, logically, a second goal is not more valuable than a first goal without which there wouldn't be a second goal. One more thing: of course, these are

not just the individual players' contributions – they're really the teams' goals' contributions – since no one player, save perhaps Lionel Messi or Cristiano Ronaldo – can score unaided.

4. But remember that football isn't linear; because goals are rare, first and second goals count for a whole lot more than third and fourth goals. And remember too that averages can be deceiving: a thirty-eight-goal total achieved by scoring a single goal in every match will produce far more points than two matches with six goals in each, five with a brace, sixteen with a single goal and fifteen with none.
5. Galeano (2003), p. 209.
6. Ibid., p. 1.

Chapter 4. Light and Dark

1. http://soccernet.espn.go.com/columns/story?id=372107&root=worldcup&cc=5739
2. Wilson (2009), p. 324. There are wonderful echoes, echoes that Menotti was probably quite aware of, of the dismissive sentiments expressed by Jorge Luis Borges, Argentina's foremost writer and intellectual: 'el fútbol es popular porque la estupidez es popular' – football is popular because stupidity is popular.
3. Wilson (2009), p. 324.
4. The scientific approach, also called logical positivism and closely associated with the thought of Karl Popper, generally is about disconfirming hypotheses – showing things to be wrong – rather than confirming them.
5. To allow for a fair comparison, we needed the same number of clubs (and possible points earned) each year. We chose to analyse data only from 2000 because the number of clubs in the Premier League fluctuated before this period.

6. Sid Lowe and Dominic Fifield, 'Chelsea Play With Fear and Lack Courage, Claims Barcelona's Dani Alves', *Guardian*, 17 April 2012; http://m.guardian.co.uk/ms/p/gnm/op/sipkZkk3hVpje3RtcOrYA_w/view.m?id=15&gid=football/2012/apr/16/dani-alves-chelsea-barcelona-fear&cat=football
7. Chris Mooney, 'What Is Motivated Reasoning? How Does It Work? Dan Kahan Answers', *Discover Magazine*, 5 May 2011; http://blogs.discovermagazine.com/intersection/2011/05/05/what-is-motivated-reasoning-how-does-it-work-dan-kahan-answers/
8. This tendency was recognized some 400 years ago by Sir Francis Bacon: 'the human understanding, once it has adopted an opinion, collects any instances that confirm it, and though the contrary instances may be more numerous and more weighty, it either does not notice them or else rejects them, in order that this opinion will remain unshaken'. This particular psychological predisposition is called 'confirmation bias'. See Bacon (1994 [1620]), p. 57.
9. Gilovich, Vallone and Tversky (1985).
10. Samuel McNerney, 'Cognitive Biases in Sports: The Irrationality of Coaches, Commentators and Fans', *Scientific American*, 22 September 2011; http://blogs.scientificamerican.com/guest-blog/2011/09/22/cognitive-biases-in-sports-the-irrationality-of-coaches-commentators-and-fans/
11. Simon Kuper, 'A Football Revolution', *Financial Times*, 17 June 2011; www.ft.com/cms/s/2/9471db52-97bb-11e0-9c37-00144feab49a.html#ixzz1tAmHkURd
12. Hearst (1991). See also Kelley (1972).
13. Dunning and Parpal (1989).
14. The Secret Footballer (2012), p. 91.
15. For the original study, see Treisman and Souther (1985).
16. Wilson, Wood and Vine (2009).
17. Binsch et al. (2010). See also Wegner (2009).

18. David Staples, 'A Quick Conversation with Bill James, the Baseball Stats King, about Hockey Stats', *Edmonton Journal*, 30 March 2012; http://blogs.edmontonjournal.com/2012/03/20/a-quick-conversation-with-bill-james-the-baseball-stats-king-about-hockey-stats/
19. Quoted in Jonathan Wilson, 'Get-Well Wishes to Argentina's El Flaco Whose Football Moved the World', *Guardian*, 16 March 2011; www.guardian.co.uk/football/blog/2011/mar/16/cesar-luis-menotti-argentina

Chapter 5. Piggy in the Middle

1. Hesse-Lichtenberger (2003), p. 113.
2. Jamie Jackson, 'Arsène Wenger Keeps Faith with Arsenal's Training Methods', *Guardian*, 23 September 2011; www.guardian.co.uk/football/blog/2011/sep/23/arsene-wenger-arsenal-training
3. 29,688 to be exact.
4. Only two teams had less possession in a match the entire season: West Brom had 26.1% possession at home against Arsenal, and Blackburn managed the league's possession low that year at 24.5% at home against Manchester United.
5. The precise number is 11.393 km; Di Salvo et al. (2007).
6. Carling (2010). Other academic research has shown that more successful teams cover more distance at high speeds than the less successful teams (Rampinini et al., 2009), and the number of actions with the ball has increased over time (Di Salvo et al., 2007; Williams, Lee and Reilly, 1999).
7. Our own analyses of Opta Sports data for three Premier League seasons (2008/09–2010/11) show that players individually had forty-two possessions per ninety minutes.

8. Brian Burke, 'Super Bowl XLII and Team Possessions', 3 February 2008; www.advancednflstats.com/2008/02/super-bowl-xlii-and-team-possessions.html
9. www.teamrankings.com/nba/stat/possessions-per-game
10. This number includes all possession changes due to unsuccessful passes (including flick-ons and lay-offs), throw-ins to an opposition player, shots off target, corners conceded, fouls conceded, unsuccessful dribbles, goals, lost tackles and dispossessions.
11. Excluding blocked shots.
12. Jaeson Rosenfeld, 'Why Players, Teams Are Undifferentiated On "Passing Skill" ', StatDNA.com, 4 May 2011; http://blog.statdna.com/post/2011/05/04/-Differentiation-in-passing-skill-between-players-and-teams-is-non-existent.aspx
13. The relationship also is very slightly curvilinear, with the gradient increasing more rapidly for teams completing more than 70% of passes in a match. This means that teams that are utterly dominant accumulate pass volume more quickly.
14. Hughes, Robertson and Nicholson (1988).
15. Hughes and Churchill (2004).
16. Lawlor et al. (2004).
17. Jones, James and Mellalieu (2004).
18. James Lawton, 'Barcelona Legend Johan Cruyff at 65 Can Still Teach the Magical Lionel Messi a Thing or Two', *Independent*, 26 April 2012; www.independent.co.uk/sport/football/news-and-comment/james-lawton-barcelona-legend-johan-cruyff-at-65-can-still-teach-the-magical-lionel-messi-a-thing-or-two-7679141.html
19. We calculated the outcomes for teams that had more possession than their opponents, that passed more accurately and played more passes than the league median (50% of the league), and that turned the ball over less to the other side in a match. We then

compared the records of these teams to those that were less successful on these performance indicators. Possession is measured as the possession percentage; pass accuracy is the percentage of completed passes; the number of passes is a simple count of passes; turnover (strict) is the percentage of recoveries given away in the match relative to the opponent (as defined earlier in the chapter); and turnover (loose) is the percentage of all lost balls committed by a team.

20. The differences are 1.47 to 1.1 (scored) and 1.15 to 1.54 (conceded) for the stricter definition of turnovers and 1.44 to 1.13 (scored) and 1.19 to 1.49 (conceded) for the more inclusive one.
21. For ease of interpretation, we defined possession percentage as Opta Sports do by the relative percentage of passes in a match.

Chapter 6. The Deflation of the Long Ball

1. Simon Kuper, 'England's Overachieving Managers', *Financial Times*, 28 January 2012; www.ft.com/cms/s/2/8de918-481d-11e1-14-00144feabdco.html#axzz1uNtBtvpK
2. Tomkins, Riley and Fulcher (2010), p. 23, compares all managers who managed at least two seasons in the Premier League.
3. The precise numbers are as follows. Premier League: 62.39 minutes, La Liga: 61.48 minutes, Bundesliga: 61.22 minutes, Serie A: 65.15 minutes. Paul Doyle, 'Number-crunching Makes Grim Reading for Arsenal's Defence', *Guardian*, 26 May 2011; www.guardian.co.uk/football/blog/2011/may/26/premier-league-opta-statistics
4. Hughes and Franks (2005).
5. Notice the major outlier on the chart – Wigan. We'll come back to them in the next chapter.

Chapter 7. Guerrilla Football

1. More specifically, Kuper and Szymanski (2009) studied the wage spending of forty English football clubs (relative to the average club) across the top two divisions of English football between 1978 and 1997 to see how well spending on wages can explain (in a statistical sense) differences in league position. According to their analyses, wages explain 92% of that variation.
2. A good technical discussion of the importance of money – wages and transfers – and on-field success can be found on the excellent Transfer Price Index website and the associated analysis; e.g., see Zach Slaton, 'A Comprehensive Model for Evaluating Total Team Valuation (TTV)'; http://transferpriceindex.com/2012/05/a-comprehensive-model-for-evaluating-total-team-valuation-ttv/
3. http://swissramble.blogspot.co.uk/2011/06/wigan-athletics-unlikely-survival.html
4. When we estimate a logistic regression of the odds of relegation using twenty years' worth of financial data and a club's wage spend relative to the average in the league, and then plug the odds for Wigan into the equation calculating the cumulative odds of relegation over five years, we find that Wigan's cumulative odds of being relegated over the course of the five years are closer to 99%, and those of Manchester United are essentially zero.
5. Opta data show that the Latics were among the league leaders for making passes in their own defensive third of the pitch – as frequently as some of the high-possession teams like Arsenal. Of course, they didn't play nearly as many passes as the Gunners in front of the other teams' goals.
6. A 'fast break' is defined by Opta as follows: 'If possession is won in the team's defensive half and within two passes the team is in

the final third [of the opposition team's half] and there is a shot, the pattern of play is logged as a fast break (counter attack)'.

7. www.zonalmarking.net/2012/05/16/wigan-stay-up-after-a-switch-to-3-4-3/
8. Vialli and Marcotti (2006), p. 136.
9. Malcolm Gladwell, 'How David Beats Goliath – When Underdogs Break the Rules', *The New Yorker*, 11 May 2009; www.newyorker.com/reporting/2009/05/11/090511fa_fact_gladwell
10. Ibid.
11. David Whitley, '"Perception of Craziness" or Flat-out Football Genius?', *Sporting News*, 30 September 2011; http://aol.sportingnews.com/ncaa-football/story/2011-09-30/perception-of-craziness-or-flat-out-football-genius. See also Romer (2006).
12. Jeff Ma, 'Belichick Was Right'; www.huffingtonpost.com/jeff-ma/belichick-was-right_b_358653.html

Chapter 8. O! Why a Football Team Is Like the Space Shuttle

1. Cited by Simon Kuper, 'A Football Revolution', *Financial Times*, 17 June 2011; www.ft.com/cms/s/2/9471db52-97bb-11e0-9c37-00144feab49a.html
2. Chris Ryan, 'Upstairs, Downstairs: Riding the Roller Coaster of Promotion and Relegation in English Football'; www.grantland.com/story/_/id/6744366/upstairs-downstairs
3. Jacob Steinberg, live match report, *Guardian*, 30 May 2011; www.guardian.co.uk/football/2011/may/30/reading-swansea-championship-play-off-live
4. We are assuming that the average talent levels are roughly similar. If we pulled Carles Puyol out of Barcelona's back line and,

with a nod to Monty Python's famous skit, substituted the real Socrates – the inquisitive Greek and not the brilliant Brazilian – in a match against Spanish third-tier team, Melilla, Socrates would be the worst player on the pitch but the Blaugrana would be likely to cruise to victory. However, this philosophically weakened Barcelona might well struggle against a second-tier club such as Girona whose talents begin to approach those of the big club and who would question Socrates relentlessly down the right side of the offensive half.

5. Kremer (1993), pp. 557, 551. One of the more rigorous applications of the O-ring theory is a recent paper by a pair of economists examining changes in the size, operations and wages of hog farms (see Yu and Orazem, 2011). The weak links in hog farms include viruses, infections and waste disposal.
6. Alex Miller and Nick Harris, 'REVEALED: Official English Football Wage Figures for the Past 25 Years'; www.sportingintelligence.com/2011/10/30/revealed-official-english-football-wage-figures-for-the-past-25-years-301002/
7. The precise multipliers are 1.9, 2.9 and 5.5.
8. We made that up. Actually they have a Physiotherapist, too.
9. Honigstein (2008), pp. 55–6.
10. Vöpel (2006).
11. For more information on the rankings see www.castrolfootball.com/rankings/rankings/
12. MIT Sports Analytics Conference, 2011; www.sloansportsconference.com/?p=626
13. Because the Castrol ratings are based on an evaluation of on-field events – touches, tackles, passes, etc. – during actual games, they contain a couple of biases. First, those players on the tail-end of a club's roster who receive little to no playing time are assigned very low scores. Presumably, the relative talent levels of the

eighth and fourth midfielders – a rare substitute and steady starter, respectively – on Lazio's 2010/11 roster are much closer than the ratio of their Castrol points (36/555 = 0.065). Cristian Brocchi, in this example, cannot have fifteen times more talent or be fifteen times more inherently productive than Pasquale Foggia. Accordingly, the ratings are really only valid for those players who appear on the pitch with some regularity, and we have designated a team's eleventh-ranked player as its weak link. The second bias concerns the overrating of goalkeepers. For many weaker teams, the player with the most Castrol points, sometimes by a wide margin, is their keeper. For example, Chievo Verona finished in the bottom half of the Serie A table in 2010/11, and their goalkeeper, Stefano Sorrentino, had 737 Castrol points, while their second-ranked player, forward Sergio Pellissier, was scored much lower at 595 points. In fact, those clubs in Europe whose best player, according to Castrol, was a netminder were outscored by an average of ten and a half goals while those whose best player was not a keeper outscored their opponents by an average of three and a quarter goals. So, for the analyses that follow we will designate as a club's strongest link their most highly rated field player. No goalkeepers, much as it pains Chris to admit this, allowed.

14. Wilson (2009), pp. 347–8, emphasis added.
15. Ibid., p. 314. This anecdote begs a lot of questions of course: really, never scored once, even on a deflection? Why didn't Gullit and van Basten organize their side? Shouldn't they have had a few guys sit on the sidelines so as to leave more space on the half-pitch and then organize midfield and attack lines? Isn't Sacchi really saying that a defence with superior players and with outstanding synchronization, that is, no weak links, will beat the most talented and advantaged offence? This exercise is also the

ultimate demonstration that possession in and of itself does not win matches.

16. Ibid., p. 236, emphasis added.
17. 'The Godfather of Models'; www.komkon.org/~ps/DK/zelen.html
18. One explanation for the big gap between Messi's club and national team performances is that the weakest links for Barcelona (the most error-prone player, the least reliable connection and understanding between two players) are very much stronger than the weakest links for Argentina. There is an interaction between the numbers for all players, including the strongest and weakest links. Here, in the interest of keeping the analysis relatively simple, we are not accounting for this additional interaction and multiplying.
19. Quantitatively able readers will note that the Messi problem is that of a significant outlier, and that we could also solve the problem by simply using ranks. With this method, the gap between Messi and Benzema (one step) is the same as that between Hummels and Piqué. The cost of using ranks is that in the middle of the talent distribution, where most players reside, ranks create relatively large differences where raw scores are quite close. Moreover, the regressions that follow look quite similar if ranks are substituted for relative quality.
20. We conducted linear regression analyses of clubs' goal differences and points, using strong- and weak-link scores as independent variables. The regression employs robust standard errors using the Castrol Edge rankings for the 2010/11 season. We control for differences across leagues by including dummy variables for four of the five leagues.
21. In our data the correlation is 0.57, which means that one explains about a third of the variation in the other.

Chapter 9. How Do You Solve a Problem Like Megrelishvili?

1. Excluding goalies, whose possession of the ball while they wave their forwards forward or gripe to their defenders about their defending and the insult of having to make a save, may sometimes take a few per cent of the game.
2. Mathematically, with the multiplicative production function, the total might increase because you're no longer multiplying the product of the qualities of the best ten players by a small fraction such as 65% or 43% that represents the weak link's efficacy.
3. Robin van Persie's red card for shooting after the whistle in Arsenal's Champions League match against Barcelona in 2010, a whistle he might not have heard over the 95,000 screaming fans, stands as a stark counter-example. There are, of course, others, but it remains true that the worst players attract a bigger share of dismissals.
4. See Jaeson Rosenfeld of StatDNA for a similar conclusion using statistics from Brazilian Serie A: http://blog.statdna.com/post/2011/03/18/Impact-of-Red-Cards-on-net-goals-and-standings-points.aspx
5. For the statistics-enamoured: we ran an ordered logit regression with home advantage, shots, goals, fouls against and red cards received in a match, along with dummy variables for the leagues (using the Premier League as the residual category).
6. Other analysts have come to a similar conclusion about the effects of red cards. Jan Vecer, Frantisek Kopriva and Tomoyuki Ichiba of Columbia University, for example, examined the effect of red cards on match outcomes in the 2006 World Cup and the 2008 Euros. Using the instantaneous shift in the betting markets

when one side receives a red card, they found that the scoring intensity of the penalized team falls by a third while the scoring intensity of the opponent rises by a quarter (Vecer, Kopriva and Ichiba, 2009). Another statistical paper based on data from the Bundesliga showed that a red card cost the home team 0.30 expected points, while its effects on the visiting team depended on the time of the card – at the thirtieth minute, the visitors lose almost half an expected point, but if the red card is awarded after the seventieth minute, they can play defensively and, on average, get away with losing no points (Mechtel et al., 2010).

7. Jonathan Wilson, 'The Question: Why is Full-back the Most Important Position on the Pitch?', *Guardian*, 25 March 2009; www.guardian.co.uk/football/blog/2009/mar/25/the-question-full-backs-football
8. Kuper (2011), p. 69.
9. Michael Cox, 'Did Manchester United Deliberately Target Gael Clichy?'; www.zonalmarking.net/2010/02/02/did-manchester-united-deliberately-target-gael-clichy/
10. Phil McNulty, BBC match report; http://news.bbc.co.uk/sport2/hi/football/eng_prem/8485984.stm
11. Wilson (2009), p. 312.
12. Andy Brassell, 'Retro Ramble: AC Milan 5 Real Madrid 0, 19th April 1989'; www.thefootballramble.com/blog/entry/retro-ramble-ac-milan-5-real-madrid-0-19th-april-1989
13. Goldblatt (2008), p. 432.
14. Quoted Wilson (2009), p. 173.
15. Del Corral, Barros and Prieto-Rodriguez (2008).
16. Myers (2012).
17. There are a few other conditions limiting this optimal substitution algorithm – no red cards for either side, no injury replacements and no extra time.
18. Carling et al. (2010).

19. Carling and Bloomfield (2010).
20. Carling et al. (2010), p. 253.
21. Henle (1978).
22. Kerr and Hertel (2011).
23. Pat Forde, '"No Way" Turns into "No Quit" for Lezak, Men's Relay Team'; http://sports.espn.go.com/oly/summer08/columns/story?columnist=forde_pat&id=3529125
24. Lisa Dillman, 'A Team Player Who Rises to the Challenge', *Los Angeles Times*, 12 August 2008; http://articles.latimes.com/2008/aug/12/sports/sp-olylezak12
25. Hüffmeier and Hertel (2011).
26. In the business world, supermarket cashiers have manifested the Köhler effect due to social comparison when a more productive cashier takes the helm in a check-out lane near them (Mas and Moretti, 2009).
27. Allen Iverson news conference transcript, 10 May 2002; http://sportsillustrated.cnn.com/basketball/news/2002/05/09/iverson_transcript/
28. Quoted in Grant Wahl, 'The World's Team', *Sports Illustrated*, 8 October 2012.
29. This section is based on the fabulous work and paper by Hamilton, Nickerson and Owan (2003).
30. In keeping with standard economic models, Koret management believed that a piece-rate system would generate maximum effort from each worker, which would then translate into the fastest possible stitching. If each stitch in a skirt was done as quickly as possible, then surely the piece-rate system was maximizing productivity. It's classic Adam Smith/Charlie Chaplin labour economics: thorough division of labour down to the narrowest task (Charlie twisting two bolts), and then monetary incentives to focus the mind and produce the effort (1/100¢ per twist).
31. Hamilton, Nickerson and Owan (2003), pp. 468–9.

32. Berg et al. (1996).
33. Franck and Nüesch (2010).
34. Ibid., pp. 220–21.

Chapter 10. Stuffed Teddy Bears

1. 'England's Longest Serving Football Manager'; www.bbc.co.uk/liverpool/content/articles/2008/11/06/north_west_football_manager_s14_w8_feature.shtml
2. We will do this at a broad enough level that allows us to encompass both managers who act like head coaches and those who are delegators and administrators.
3. Carlyle (1840), pp. 1–2.
4. Ibid., pp. 12–13.
5. Ibid., p. 13.
6. Ronay (2010), Introduction.
7. Cited in Ken Jones, 'Searching for the Secrets of Shankly', *Independent*, 23 February 1996; www.independent.co.uk/sport/searching-for-the-secrets-of-shankly-1320604.html
8. Carlyle (1840), p. 17.
9. Barney Ronay, 'Football Managers: Camel Coat Optional', *Guardian*, 12 August 2009; www.guardian.co.uk/football/2009/aug/12/football-managers-leadership
10. Or, even any gargantuan critic – Sam Allardyce, bemoaning his fate to lead mainly small, under-resourced clubs said in the autumn of 2010, 'I'm not suited to Bolton or Blackburn, I would be more suited to Inter or Real Madrid. It wouldn't be a problem to me to go and manage those clubs because I would win the double or the league every time. Give me Manchester United or Chelsea and I would do the same, it wouldn't be a problem.' Tomkins, Riley and Fulcher (2010), p. 35.

11. This means that the statistical model's so-called r-squared was 0.89. Kuper and Szymanski (2009).
12. Ibid., p. 111.
13. Kuper and Szymanski (2009) disagree; they argue that the market for managers is inefficient. Therefore, the amount of variation in league position explained by the wage data reflects player talent, not managerial talent. After all, if there is more randomness in good managers ending up with good clubs, then the market for managers should be less efficient.
14. There is a strong correlation between a club's transfer spend and spending on wages. See Zach Slaton, 'A Comprehensive Model for Evaluating Total Team Valuation (TTV)'; http://transferpriceindex.com/2012/05/a-comprehensive-model-for-evaluating-total-team-valuation-ttv/
15. Kuper and Szymanski (2009) say that the 'correlation' drops to 70%. This statement is slightly confusing to the statisticians in us, as correlations are typically not expressed in percentage terms. A correlation of 0.7 would translate to 49% of the variation in league position explained by wages year on year.
16. Bridgewater (2010).
17. The debate over whether football clubs are utility- or profit-maximizing organizations is several decades old. For a nice overview of the various issues involved, see Szymanski and Kuypers (1999), or Szymanski (2009).
18. Heuer et al. (2011).
19. For the statistically minded, the question is this: do you compare the significant coefficients on the CEO fixed-effects variables to those of industry and firm in the overall regression or to the size of the residual?
20. Keri (2011), p. 13.
21. Ibid., p. 119.

22. A fine example of the mechanical is this passage from Dawson, Dobson and Gerrard (2000), p. 401:

    Firms are organizational and technical units for the production of commodities. The production process involves the transformation of factor inputs into product outputs. The production function represents the technical relationship between inputs and outputs. As a theoretical concept the production function is usually based on the assumption of full technical efficiency such that output is maximized for any given level of inputs (or inputs are minimized for any given level of output).

23. The logic is doubtful: we find it hard to believe that doubling or tripling the wages of Wigan's squad will turn them into Champions League contenders.
24. Bennedsen, Pérez-González and Wolfenzon (2010).

Chapter 11. The Young Prince

1. Gabriele Marcotti, 'Meet Portugal's Boy Genius', *Wall Street Journal*, 5 October 2010; http://online.wsj.com/article/SB10001424052748704380504575530111481441870.html
2. Katy Murrells, 'Chelsea an "Embarrassment" to League and Next Manager Faces "Hell"', *Guardian*, 5 March 2012; www.guardian.co.uk/football/2012/mar/05/chelsea-manager-scolari-villas-boas
3. Keri (2011), p. 101.
4. As quoted in Vialli and Marcotti (2006), p. 121.
5. The study sought to explain how a club's table position was affected by manager characteristics and club wages. Bridgewater, Kahn and Goodall (2009).
6. As quoted in Vialli and Marcotti (2006), p. 115.

7. Bridgewater Kahn and Goodall (2009), p. 17.
8. Michaels, Handfield-Jones and Axelrod (2001), p. 101.
9. Ibid.
10. It's an amusing image though – the Newborn World Cup. You can almost hear Alan Smith criticizing the neonate for an awfully heavy and awkward first touch. The media pile on: all England ever produce are babies who play inelegant route one 4–4–2 football, choke in the big competitions, are terrible at penalties and dirty their nappies.
11. Sloboda et al. (1996).
12. Duncan White, 'André Villas-Boas: Chelsea's New Manager Who Has Dedicated Himself to Football', *Telegraph*, 25 June 2011; www.telegraph.co.uk/sport/football/teams/chelsea/8597265/Andre-Villas-Boas-Chelseas-new-manager-who-has-dedicated-himself-to-football.html
13. 'Terry Venables: Fernando Torres Needs to Forget about Scoring', *Sun*, 16 September 2011; www.thesun.co.uk/sol/homepage/sport/football/3819274/Terry-Venables-Fernando-Torres-needs-to-forget-about-scoring.html?OTC-RSS&ATTR=Football
14. Groysberg (2010), p. 40.
15. This is a larger commitment for a club or organization than just refusing to interview outside candidates. The club must put systems in place and dedicate resources to grooming and developing assistants and making them ready for promotion when the time comes. American sports franchises are much more successful in fostering the talents of their assistant coaches and preparing them for the job of head coach than are European football clubs.
16. Didier Drogba, *C'était Pas Gagné*, Paris: Editions Prologations, 2008. Quoted in Kuper and Szymanski (2012), p. 31.
17. A figure of 40% hits hasn't been seen in Major League Baseball since Ted Williams had a batting average of .406 in 1941.

18. Data from Nate Silver on the Baseball Prospectus website: www.baseballprospectus.com/article.php?articleid=1897
19. For Norway see Arnulf, Mathisen and Haerem (2012); for Germany see Heuer et al. (2011); for Italy see De Paola and Scoppa (2009); and for England see Dobson and Goddard (2011).
20. ter Weel (2011).
21. Ben Webster, 'Speed Camera Benefits Overrated', *The Times*, 16 December 2005.
22. www.nobelprize.org/nobel_prizes/economics/laureates/2002/kahneman-autobio.html
23. Kelly and Waddington (2006).
24. Okwonga (2010), pp. 138–9.
25. Gallimore and Tharp (2004); Becker and Wrisberg (2008).
26. Quoted in Kelly and Waddington (2006).
27. Quoted in Ericsson, Prietula and Cokely (2007).

Chapter 12. Life During the Reformation

1. Simon Burnton, 'Billy Beane Leaves Moneyball Behind to Refocus on Statistical Truths', *Guardian*, 22 March 2012; www.guardian.co.uk/sport/2012/mar/22/billy-beane-moneyball
2. Simon Kuper, 'Sky-blue Thinking', *Financial Times*, 27 August 2012; www.ft.com/cms/s/2/04e0e834-e6c5-11e1-af33-00144feab49a.html#axzz26iYkWJez
3. Steve Lohr, 'The Age of Big Data', *The New York Times*, 11 February 2012; www.nytimes.com/2012/02/12/sunday-review/big-datas-impact-in-the-world.html?pagewanted=all&_r=0
4. Barney Ronay, 'Euro 2012: Valeriy Lobanovsky, King of Kiev Who Was Before His Time', *Guardian*, 15 June 2012; www.guardian.co.uk/football/blog/2012/jun/15/euro-2012-valeriy-lobanovsky-kiev
5. Grund (2012).

6. Tomkins, Riley and Fulcher (2010), p. 56.
7. Lothar Gorris and Thomas Hüetlin, 'The Pitch Is a Jungle', *Der Spiegel*, 30 June 2006; www.spiegel.de/international/spiegel/interview-with-football-philosopher-jorge-valdano-the-pitch-is-a-jungle-a-424493-2.html
8. The part about untested players being Moneyball-like, the part about paying a lot not.

Extra Time – The Numbers Game at the World Cup

1. From an interview with Chumpitaz in Keme Nzerem, 'Henry Kissinger and Football's Longest Unsolved Riddle', 4 April 2012; http://www.channel4.com/news/dr-henry-kissinger-and-footballs-longest-unsolved-riddle
2. Quoted in David Yallop (2011).
3. Quoted in Goldblatt (2008).
4. Analytical readers will note, correctly, that this is a correlation and that causality could run in either direction. A huge victory could give the players confidence and boost their performance for the rest of the tournament, or a large margin could reflect the excellent form of the squad and their inability to downshift their own performance level.
5. Jeff Z. Klein, 'FIFA and the Drawing of Lots: A Brief History', 19 June 2010, Goal: *New York Times* Soccer Blog; http://goal.blogs.nytimes.com/2010/06/19/fifa-and-the-drawing-of-lots-a-brief-history
6. Adrian Escude, quoted in Max Grieve, 'A Blindfolded Boy with His Hand in a Pot: In Favour of Penalty Shoot-outs', 2013; http://afootballreport.com/post/23921754762/a-blindfolded-boy-with-his-hand-in-a-pot-in-favour-of
7. This research and more can be found in his excellent 2014 book, *Beautiful Game Theory*, Princeton: Princeton University Press.

8. Palacios-Huerta (2014) offers a very clever solution by reordering the kicks from the seemingly fair, Uruguay-Ghana-U-G-U-G- . . . to a more tennis-like, flip-flopping pattern: U-G-G-U-G-U-U-G-G-U-U-G-U-G-G-U- . . .
9. Steven Gerrard (2006), pp. 464–5.
10. Philip Furley, Matt Dicks, Fabian Stendtke and Daniel Memmert (2012), '"Get it out the way. The wait's killing me": Hastening and Hiding During Soccer Penalty Kicks', *Psychology of Sport and Exercise* 13 (4): 454–65.
11. R. S. W. Masters, J. van der Kamp and R. C. Jackson (2007), 'Imperceptibly Off-Center Goalkeepers Influence Penalty-Kick Direction in Soccer', *Psychological Science* 18 (3): 222–3.
12. John van der Kamp and Rich S. W. Masters (2008), 'The Human Müller-Lyer Illusion in Goalkeeping', *Perception* 37 (7): 951–4.
13. Greg Wood and Mark R. Wilson (2010), 'A Moving Goalkeeper Distracts Penalty-takers and Impairs Shooting Accuracy', *Journal of Sports Sciences* 28 (9): 937–46.
14. Martina Navarro, John van der Kamp, Ronald Ranvaud and Geert J. P. Savelsbergh (2013), 'The Mere Presence of a Goalkeeper Affects the Accuracy of Penalty Kicks', *Journal of Sports Sciences* 31 (9): 921–9.
15. Frank C. Bakker, Raôul R. D. Oudejans, Olaf Binsch and John van der Kamp (2006), 'Penalty Shooting and Gaze Behavior: Unwanted Effects of the Wish Not to Miss', *International Journal of Sport Psychology* 37 (2–3): 265–80; Mark R. Wilson, Greg Wood and Samuel J. Vine (2009), 'Anxiety, Attentional Control, and Performance Impairment in Penalty Kicks', *Journal of Sport and Exercise Psychology* 31 (6): 761–75; Benjamin Noël and John van der Kamp (2012), 'Gaze Behaviour During the Soccer Penalty Kick: An Investigation of the Effects of Strategy and Anxiety', *International Journal of Sport Psychology* 43(4): 1–20; Raôul R. D. Oudejans, Olaf Binsch and Frank C. Bakker (2013), 'Negative Instructions and Choking

under Pressure in Aiming at a Far Target,' *International Journal of Sport Psychology* 44 (4): 294–309.

16. Geir Jordet, Esther Hartman and Pieter Jelle Vuijk (2012), 'Team History and Choking under Pressure in Major Soccer Penalty Shootouts', *British Journal of Psychology* 103(2): 268–83.
17. 'England: Roy Hodgson May Use Psychologist for Penalties', BBC Sport websites, 25 February 2014; http://www.bbc.com/sport/0/football/26332893

Bibliography

Arnulf, Jan Ketil, John Erik Mathisen and Thorvald Haerem (2012), 'Heroic Leadership Illusions in Football Teams: Rationality, Decision Making and Noise Signal Ratio in the Firing of Football Managers', *Leadership* 8 (2): 169–85

Ayton, Peter and Anna Braennberg (2008), 'Footballers' Fallacies', in Patric Andersson, Peter Ayton and Carsten Schmidt (eds.), *Myths and Facts About Football: The Economics and Psychology of the World's Greatest Sport*. Newcastle upon Tyne: Cambridge Scholars Publishing

Bacon, Francis (1994 [1620]), *Novum Organum*, trans. Peter Urbach and John Gibson. Chicago: Open Court Publishing

Becker, A. J. and C. A. Wrisberg (2008), 'Effective Coaching in Action: Observations of Legendary Collegiate Basketball Coach Pat Summitt', *The Sport Psychologist* 22 (2): 197–211

Ben-Naim, Eli, Federico Vazquez and Sidney Redner (2006), 'Parity and Predictability of Competitions', *Journal of Quantitative Analysis in Sports* 2 (4); physics.bu.edu/~redner/pubs/pdf/jqas.pdf

Bennedsen, Morten, Francisco Pérez-González and Daniel Wolfenzon (2010), 'Do CEOs Matter?', Working Paper, INSEAD

Berg, Peter, Eileen Appelbaum, Thomas Bailey and Arne L. Kalleberg (1996), 'The Performance Effects of Modular Production in the Apparel Industry', *Industrial Relations* 35 (3): 356–73

Binsch, Olaf, Raôul R. D. Oudejans, Frank C. Bakker and Geert J. P. Savelsbergh (2010), 'Ironic Effects and Final Target Fixation in a Penalty Shooting Task', *Human Movement Science* 29 (2): 277–88

Bortkiewicz, Ladislaus von (1898), *Das Gesetz der kleinen Zahlen*. Leipzig: E. G. Teubner

Bridgewater, Sue (2010), *Football Management*. London: Palgrave Macmillan

Bridgewater, Sue, Lawrence M. Kahn and Amanda H. Goodall (2009), 'Substitution Between Managers and Subordinates: Evidence from British Football', NCER Working Paper Series 51, National Centre for Econometric Research

Brillinger, David R. (2010), 'Soccer/World Football', in James J. Cochran (ed.), *Encyclopedia of Operations Research and Management Science*. New York: Wiley

Brocas, Isabelle and Juan D. Carrillo (2004), 'Do the "Three-Point Victory" and "Golden Goal" Rules Make Soccer More Exciting? A Theoretical Analysis of a Simple Game', *Journal of Sports Economics* 5 (2): 169–85

Carling, Chris (2010), 'Analysis of Physical Activity Profiles When Running with the Ball in a Professional Soccer Team', *Journal of Sports Sciences* 38 (3): 319–26

Carling, Christopher and Jonathan Bloomfield (2010), 'The Effect of an Early Dismissal on Player Work-rate in a Professional Soccer Match', *Journal of Science and Medicine in Sport* 13 (1): 126–8

Carling, Christopher, Vincent Espié, Franck Le Gall, Jonathan Bloomfield and Hugues Jullien (2010), 'Work-rate of Substitutes in Elite Soccer: A Preliminary Study', *Journal of Science and Medicine in Sport* 13 (2): 253–5

Carlyle, Thomas (1840), *On Heroes, Hero-Worship and the Heroic in History*. London: Chapman and Hall

Colvin, Geoffrey (2010), *Talent Is Overrated: What Really Separates World-Class Performers from Everyone Else.* New York: Penguin

Cullis, Stan (1961), *All for the Wolves.* London: Sportsman Book Club

Dawson, Peter, Stephen Dobson and Bill Gerrard (2000), 'Estimating Coaching Efficiency in Professional Team Sports: Evidence from English Association Football', *Scottish Journal of Political Economy* 47 (4): 399–421

De Paola, Maria and Vincenzo Scoppa (2009), 'The Effects of Managerial Turnover: Evidence from Coach Dismissals in Italian Soccer Teams', Working Paper, Department of Economics and Statistics, University of Calabria, Cosenza, Italy

del Corral, Julio, Carlos Pestana Barros and Juan Prieto-Rodriguez (2008), 'The Determinants of Soccer Player Substitutions: A Survival Analysis of the Spanish Soccer League', *Journal of Sports Economics* 9 (2): 160–72

del Corral, Julio, Juan Prieto-Rodriguez and Rob Simmons (2010), 'The Effect of Incentives on Sabotage: The Case of Spanish Football', *Journal of Sports Economics* 11 (3): 243–60

Di Salvo, V., R. Baron, H. Tschan, F. J. Calderon Montero, N. Bachl and F. Pigozzi (2007), 'Performance Characteristics According to Playing Position in Elite Soccer', *International Journal of Sports Medicine* 28 (3): 222–7

Dilger, Alexander and Hannah Geyer (2009), 'Are Three Points for a Win Really Better Than Two? A Comparison of German Soccer League and Cup Games', *Journal of Sports Economics* 10 (3): 305–18

Dobson, Stephen and John Goddard (2011), *The Economics of Football*. Cambridge: Cambridge University Press

Dunning, David and Mary Parpal (1989), 'Mental Addition Versus Mental Subtraction in Counterfactual Reasoning: On Assessing the Impact of Personal Actions and Life Events', *Journal of Personality and Social Psychology* 57 (1): 5–15

Ericsson, K. Anders, Michael J. Prietula and Edward T. Cokely (2007), 'The Making of an Expert', *Harvard Business Review*, July–August: 115–21

Franck, Egon and Stephan Nüesch (2010), 'The Effect of Talent Disparity on Team Productivity in Soccer', *Journal of Economic Psychology* 31 (2): 218–29

Galeano, Eduardo (2003), *Soccer in Sun and Shadow*. New York: Verso

Gallimore, Ronald and Roland Tharp (2004), 'What a Coach Can Teach a Teacher, 1975–2004: Reflections and Reanalysis of John Wooden's Teaching Practices', *The Sport Psychologist* 18 (2): 119–37

Gallo, Edoardo, Thomas Grund and J. James Reade (2013), 'Punishing the Foreigner: Implicit Discrimination in the Premier League Based on Oppositional Identity', *Oxford Bulletin of Economics and Statistics* 75 (1): 136–56

Garicano, Luis and Ignacio Palacios-Huerta (2005), 'Sabotage in Tournaments: Making the Beautiful Game a Bit Less Beautiful', CEPR Discussion Paper 5231

Gerrard, Steven (2006), *Gerrard: My Autobiography,* London: Transworld

Gilovich, Thomas, Robert Vallone and Amos Tversky (1985), 'The Hot Hand in Basketball: On the Misperception of Random Sequences', *Cognitive Psychology* 17 (3): 295–314

Gladwell, Malcolm (2008), *Outliers: The Story of Success*. New York: Little, Brown

Goldblatt, David (2008), *The Ball Is Round*. New York: Riverhead Books

Groysberg, Boris (2010), *Chasing Stars: The Myth of Talent and the Portability of Performance*. Princeton: Princeton University Press

Grund, Thomas U. (2012), 'Network Structure and Team Performance: The Case of English Premier League Soccer Teams', *Social Networks*, September

Hamilton, Barton H., Jack A. Nickerson and Hideo Owan (2003), 'Team Incentives and Worker Heterogeneity: An Empirical Analysis of the Impact of Teams on Productivity and Participation', *Journal of Political Economy* 111 (3): 465–97

Hastorf, Albert H. and Hadley Cantrell, 'They Saw a Game: A Case Study', *Journal of Abnormal and Social Psychology* 49 (1): 129–34

Hearst, Eliot (1991), 'Psychology and Nothing', *American Scientist* 79 (5): 432–43

Henle, Mary (1978), 'One Man Against the Nazis – Wolfgang Köhler', *American Psychologist* 33 (1): 939–44

Hesse-Lichtenberger, Uli (2003), *Tor! The Story of German Football*. London: WSC Books

Heuer, Andreas, Christian Müller and Oliver Rubner (2010), 'Soccer: Is Scoring Goals a Predictable Poissonian Process?', *Europhysics Letters* 89, 38007; http://arxiv.org/pdf/1002.0797

Heuer, Andreas, Christian Müller, Oliver Rubner, Norbert Hagemann and Bernd Strauss (2011), 'Usefulness of Dismissing and Changing the Coach in Professional Soccer', PLoS ONE 6 (3); www.plosone.org

Honigstein, Raphael (2008), *Englischer Fussball: A German's View of our Beautiful Game*. London: Yellow Jersey

Hopcraft, Arthur (1968), *The Football Man: People and Passions in Soccer*. London: Collins

Hüffmeier, Joachim and Guido Hertel (2011), 'When the Whole Is More Than the Sum of Its Parts: Group Motivation Gains in the Wild', *Journal of Experimental Social Psychology* 47 (2): 455–9

Hughes, Charles (1990), *The Winning Formula*. London: Collins

Hughes, Mike (2003), 'Notational Analysis', in Thomas Reilly and A. Mark Williams (eds.), *Science and Soccer*, 2nd edn. London: Routledge, pp. 245–64

Hughes, Mike D. and Roger Bartlett (2002), 'The Use of Performance Indicators in Performance Analysis', *Journal of Sports Sciences* 20 (10): 739–54

Hughes, Mike D. and Steven Churchill (2004), 'Attacking Profiles of Successful and Unsuccessful Teams in Copa America 2001', *Journal of Sports Sciences* 22 (6): 505

Hughes, Mike and Ian Franks (2005), 'Analysis of Passing Sequences, Shots and Goals in Soccer', *Journal of Sports Sciences* 23 (5): 509–14

Hughes, M. D., K. Robertson and A. Nicholson (1988), 'An Analysis of 1986 World Cup of Association Football', in T. Reilly, A. Lees, K.

Davids and W. Murphy (eds.), *Science and Football*. London: E&FN Spon, pp. 363–7

Jones, P. D., N. James and S. D. Mellalieu (2004), 'Possession as a Performance Indicator in Soccer', *International Journal of Performance Analysis in Sport* 4 (1): 98–102

Kato, Takao and Pian Shu (2009), 'Peer Effects, Social Networks and Intergroup Competition in the Workplace', ASB Economics Working Paper, University of Aarhus

Kelley, Harold H. (1972), 'Attribution in Social Interaction', in E. E. Jones, D. E. Kanouse, H. H. Kelley, R. E. Nisbett, S. Valins and B. Weiner (eds.), *Attribution: Perceiving the Causes of Behavior*. Morristown, NJ: General Learning Press

Kelly, Seamus and Ivan Waddington (2006), 'Abuse, Intimidation and Violence as Aspects of Managerial Control in Professional Soccer in Britain and Ireland', *International Review for the Sociology of Sport* 41 (2): 147–64

Keri, Jonah (2011), *The Extra 2%: How Wall Street Strategies Took a Major League Baseball Team from Worst to First*. New York: Random House

Kerr, Norbert L. and Guido Hertel (2011), 'The Köhler Group Motivation Gain: How to Motivate the "Weak Links" in a Group', *Social and Personality Psychology Compass* 5 (1): 43–55

Kremer, Michael (1993), 'The O-Ring Theory of Economic Development', *Quarterly Journal of Economics* 108 (3): 551–75

Kuper, Simon (2011), *The Football Men*. London: Simon & Schuster

Kuper, Simon and Stefan Szymanski (2009), *Why England Lose: And Other Curious Phenomena Explained*. London: HarperSport

Kuper, Simon and Stefan Szymanski (2012), *Soccernomics: Why England Loses, Why Spain, Germany, and Brazil Win, and Why the US, Japan, Australia, Turkey – and Even Iraq – Are Destined to Become the Kings of the World's Most Popular Sport*. 2nd edn. New York: Nation Books

Lames, Martin (2006), 'Glücksspiel Fußball – Zufallseinflüsse beim Zustandekommen von Toren', in Barbara Halberschmidt and Bernd Strauß (eds.), *Elf Freunde sollt ihr sein!? 38. Jahrestagung der Arbeitsgemeinschaft für Sportpsychologie (asp). Abstracts*, Hamburg: Czwalina

Larsen, Oyvind (2001), 'Charles Reep: A Major Influence on British and Norwegian Football', *Soccer and Society* 2 (3): 55–78

Lawlor, J., D. Low, S. Taylor and A. M. Williams (2004), 'The FIFA World Cup 2002: An Analysis of Successful and Unsuccessful Teams', *Journal of Sports Sciences* 22 (6): 510

Lyons, Keith (1997), 'The Long and Direct Road: Charles Reep's Analysis of Association Football'; http://keithlyons.me/2011/02/28/goal-scoring-in-association-football-charles-reep/

Mas, Alexandre and Enrico Moretti (2009), 'Peers at Work', *American Economic Review* 99 (1): 112–45

McGarry, Tim and Ian M. Franks (2003), 'The Science of Match Analysis', in Thomas Reilly and A. Mark Williams (eds.), *Science and Soccer*, 2nd edn. London: Routledge, pp. 265–75

Mechtel, Mario, Tobias Brändle, Agnes Stribeck and Karin Vetter (2010), 'Red Cards: Not Such Bad News for Penalized Guest Teams', *Journal of Sports Economics* 12 (6): 621–46

Michaels, Ed, Helen Handfield-Jones and Beth Axelrod (2001), *The War for Talent*. Cambridge, MA: Harvard Business Press

Miguel, Edward, Sebastián S. M. Saiegh and Shanker Satyanath (2011), 'Civil War Exposure and Violence', *Economics & Politics* 23 (1): 59–73

Moroney, M. J. (1951), *Facts from Figures*. London: Penguin

Myers, Bret (2012), 'A Proposed Decision Rule for the Timing of Soccer Substitutions', *Journal of Quantitative Analysis in Sports* 8 (1)

Okwonga, Musa (2010), *Will You Manage? The Necessary Skills to Be a Great Gaffer*. London: Serpent's Tail

Palacios-Huerta, Ignacio (2004), 'Structural Changes During a Century of the World's Most Popular Sport', *Statistical Methods & Applications* 13 (2): 241–58

Palacios-Huerta, Ignacio (2014), *Beautiful Game Theory*. Princeton: Princeton University Press

Pollard, Richard (2002), 'Charles Reep (1904–2002): Pioneer of Notational and Performance Analysis in Football', *Journal of Sports Sciences* 20 (10): 853–5

Pollard, Richard, Jake Ensum and Samuel Taylor (2004), 'Applications of Logistic Regression to Shots at Goal in Association Football: Calculation of Shot Probabilities, Quantification of Factors and Player/Team', *Journal of Sports Sciences* 22 (6): 504

Pollard, Richard and Charles Reep (1997), 'Measuring the Effectiveness of Playing Strategies at Soccer', *Journal of the Royal Statistical Society. Series D (The Statistician)* 46 (4): 541–50

Quitzau, Jörn and Henning Vöpel (2009), 'Der Faktor Zufall im Fußball: Eine empirische Untersuchung für die Saison 2007/08', Hamburgisches WeltWirtschaftsInstitut, HWWI Kompetenzbereich Wirtschaftliche Trends Working Paper 1-22

Rampinini E., F. M. Impellizzeri, C. Castagna, A. J. Coutts and U. Wisløff (2009), 'Technical Performance During Soccer Matches of the Italian Serie A League: Effect of Fatigue and Competitive Level', *Journal of Science and Medicine in Sport* 12 (1): 227–33

Read, Brenda and Phyl Edwards (1992), *Teaching Children to Play Games*. Leeds: White Line Publishing

Reep, Charles and Bernard Benjamin (1968), 'Skill and Chance in Association Football', *Journal of the Royal Statistical Society. Series A (General)* 131 (4): 581–5

Reep, Charles, Richard Pollard and Bernard Benjamin (1971), 'Skill and Chance in Ball Games', *Journal of the Royal Statistical Society. Series A (General)* 134 (4): 623–9

Reilly, Thomas and V. Thomas (1976), 'A Motion Analysis of Work-rate in Different Positional Roles in Professional Football Match-play', *Journal of Human Movement Studies* 2: 87–97

Romer, David (2006), 'Do Firms Maximize? Evidence from Professional Football', *Journal of Political Economy* 114 (2): 340–65

Ronay, Barney (2010), *The Manager: The Absurd Ascent of the Most Important Man in Football*. London: Little, Brown

Secret Footballer, The (2012), *I Am the Secret Footballer: Lifting the Lid on the Beautiful Game*. London: Guardian Books

Skinner, Gerald K. and Guy H. Freeman (2009), 'Soccer Matches as Experiments: How Often Does the "Best" Team Win?', *Journal of Applied Statistics* 36 (10): 1087–95

Sloboda, John A., Jane W. Davidson, Michael J. A. Howe and Derek G. Moore (1996), 'The Role of Practice in the Development of Performing Musicians', *British Journal of Psychology* 87 (2): 287–309

Szymanski, Stefan (2009), *Playbooks and Checkbooks: An Introduction to the Economics of Modern Sport*. Princeton: Princeton University Press

Szymanski, Stefan and Tim Kuypers (1999), *Winners and Losers: The Business Strategy of Football*. London: Viking

ter Weel, Bas (2011), 'Does Manager Turnover Improve Firm Performance? Evidence from Dutch Soccer, 1986–2004', *De Economist* 159 (3): 279–303

Tomkins, Paul, Graeme Riley and Gary Fulcher (2010), *Pay As You Play: The True Price of Success in the Premier League Era*. Wigston: Gprf Publishing

Treisman, Anne and Janet Souther (1985), 'Search Asymmetry: A Diagnostic for Preattentive Processing of Separable Features', *Journal of Experimental Psychology: General* 114 (3): 285–310

Vecer, Jan, Frantisek Kopriva and Tomoyuki Ichiba (2009), 'Estimating the Effect of the Red Card in Soccer: When to Commit an

Offense in Exchange for Preventing a Goal Opportunity', *Journal of Quantitative Analysis in Sports* 5 (1); www.degruyter.com/view/j/jqas.2009.5.1/jqas.2009.5.1.1146/jqas.2009.5.1.1146.xml?format=INT

Vialli, Gianluca and Gabriele Marcotti (2006), *The Italian Job: A Journey to the Heart of Two Great Footballing Cultures*. London: Bantam Press

Vöpel, Henning (2006), 'Ein "ZIDANE-Clustering-Theorem" und Implikationen für den Finanzausgleich in der Bundesliga', Hamburgisches WeltWirtschaftsInstitut, HWWI Research Paper 1-3

Wegner, Daniel M. (2009), 'How to Think, Say, or Do Precisely the Worst Thing for Any Occasion', *Science* 325 (5936): 48–50

Williams A. M., D. Lee and T. Reilly (1999), *A Quantitative Analysis of Matches Played in the 1991–92 and 1997–98 Seasons*. London: The Football Association

Wilson, Jonathan (2009), *Inverting the Pyramid: The History of Football Tactics*. London: Orion

Wilson, Mark R., Greg Wood and Samuel J. Vine (2009), 'Anxiety, Attentional Control, and Performance Impairment in Penalty Kicks', *Journal of Sport & Exercise Psychology* 31 (6): 761–75

Yallop, David (2011), *How They Stole the Game*. London: Constable

Yu, Li and Peter Orazem (2011), 'O-Ring Production on U. S. Hog Farms: Joint Choices of Farm Size, Technology, and Compensation', Working Paper, Iowa State University Department of Economics, January

Index

Page references in *italic* indicate Figures or Tables.

Index

Index

Index